AMERICAN LEGENDS

BOSS
429
GOODYEAR

AMERICAN LEGENDS

MIKE MUELLER

Lowe & B. Hould
Publishers

This edition published in 2003 by Lowe & B. Hould Publishers, an imprint of Borders, Inc., 515 East Liberty, Ann Arbor, MI 48104.
Lowe & B. Hould Publishers is a trademark of Borders Properties, Inc.

First published by MBI Publishing Company, Galtier Plaza, Suite 200, 380 Jackson Street, St. Paul, MN 55101-3885 USA

ISBN 0-681-19745-5

On the front cover: (Left) Truncated two-seat renditions of Ford's mainline 1955 and 1956 models, the T-bird grew slightly in 1957 thanks to a longer tail section wearing more pronounced fins.

(Right, top) When first introduced in 1965, Shelby American's GT350 Mustang variant was a single-purpose machine available in one color only, without a back seat or an automatic transmission, and loaded with a full collection of standard race-ready hardware.

(Right, bottom) The same basic interior offered since 1958 made one last appearance in 1962.

On the title page: Hands down, the most beastly Mustang ever was the Boss 429, a hinkered-down, hairy animal with a heavy-duty lowered suspension, big 15-inch Magnum 500 wheels and 375 horses of V-8.

On the back cover: Only 300 Corvettes were built in a limited production run for 1953. Chevrolet tried again in 1954 with a nearly identical model, this time offering it in black, Pennant Red, and Sportsman Blue, along with the Polo White, the sole exterior choice the previous year.

Printed in China

Contents

Thunderbird Milestones

THE
Heartbeat
OF AMERICA

CORVETTE MILESTONES

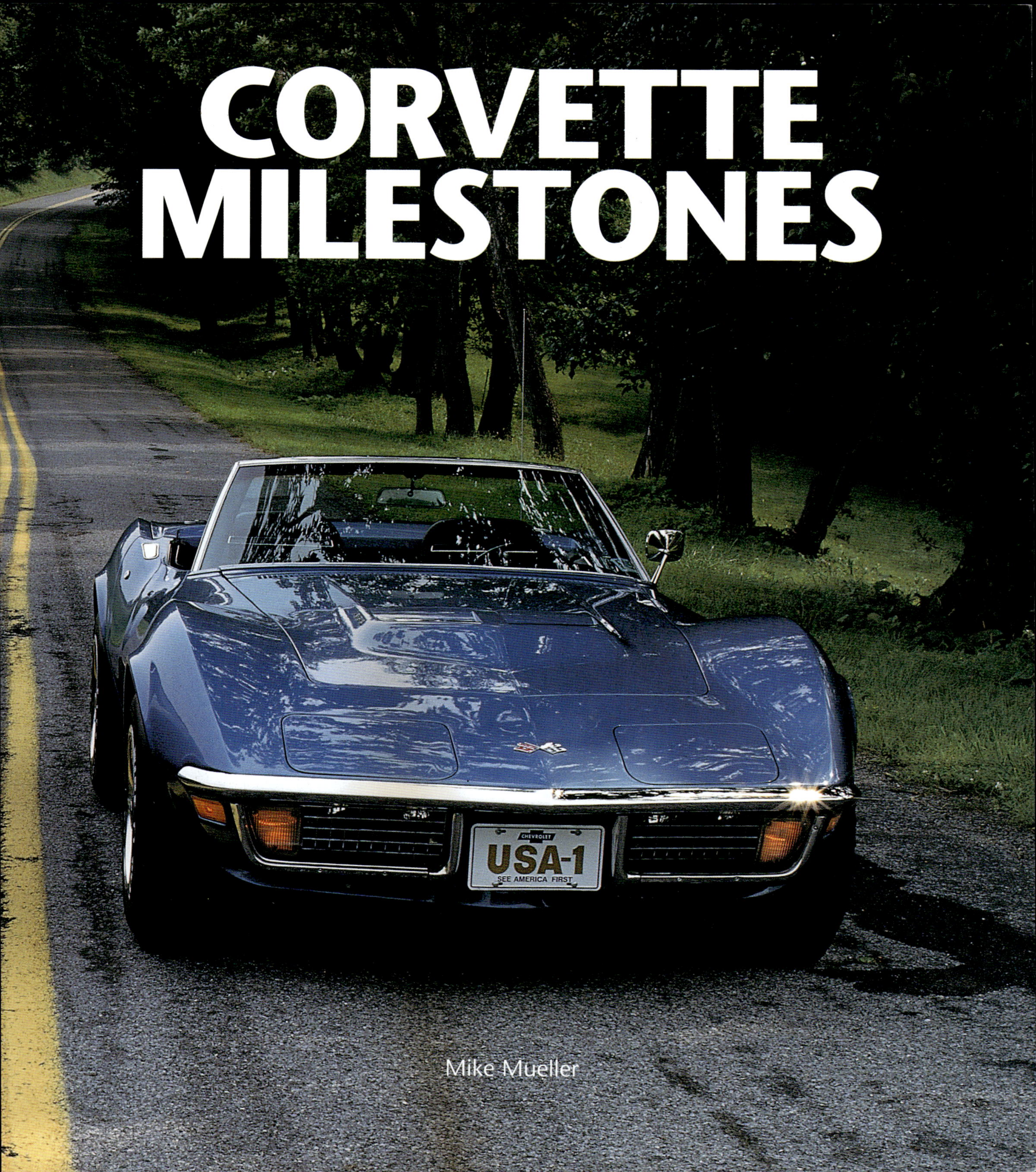

Mike Mueller

THE NATIONAL CORVETTE MUSEUM
CORVETTE

ACKNOWLEDGMENTS

It was an event some seven years in the making, or four decades depending on your perspective. Many of you out there who've lived with and loved America's sports car from its humble birth in 1953 may have wondered if the long-deserved, long-awaited National Corvette Museum would ever open its doors. Originally discussed in 1987, the idea for a four-walled tribute to one of the greatest cars this county has ever produced went through more than its fair share of ups and downs before becoming reality. But reality it finally was, right there before all our eyes. On September 2, 1994, the ribbon across the doors of Valhalla in Bowling Green, Kentucky, was finally cut.

When the National Corvette Museum finally opened in Bowling Green, Kentucky, on Labor Day weekend in 1994, the legendary piece of Corvette history that greeted visitors at the main lobby was Zora Duntov's ill-fated SS racer of 1957. Shown here in front of the museum, the magnesium-bodied SS was returned later that year to its permanent home at the Indianapolis Motor Speedway Hall of Fame museum.

Everyone was there—Zora Arkus-Duntov, Dave McLellan, Dave Hill, Jim Perkins, and Larry Shinoda. So were the Beach Boys and about 120,000 others. Nearly all the memorable cars were there as well including Duntov's 1957 SS. Mitchell's Stingray racer Manta Ray, Mako Shark, Purple People Eater, "Big Doggie," and the 1 millionth Corvette. About the only major omissions from the lineup during that first overcrowded weekend were the SR-2 and Grand Sport. But if you wanted to see the greatest collection of Corvettes ever, the National Corvette Museum's grand opening represented a once-in-a-lifetime chance to enjoy as many of them as you could possibly imagine, all together under one very high-pitched roof.

But if you were stuck at home, perhaps you could use a little sampling of what that moment was like. While this humble publication can by no means match the

Everyone was there. Zora Duntov. Dave McLellan. Dave Hill. Jim Perkins. Larry Shinoda. The Beach Boys. And about 120,000 others. Nearly all the memorable cars were there as well—Duntov's 1957 SS, Mitchell's Stingray racer, Manta Ray, Mako Shark, Purple People Eater. . . ■

grandeur of the National Corvette Museum, it can offer as much rolling legend as can fit in 96 pages. Sure, not all the greats are here, but many of them are.

Assembling this four-cornered tribute to one of the greatest cars this county has ever produced involved more than a little bit of help from countless Corvette crazies. As usual, I can't get away without mentioning good friends Greg Pernula and Paul Zazarine of Dobbs Publishing in Lakeland, Florida. Greg is editor of *Corvette Fever* and Paul is a former *CF* editor and now one of Dobbs' editorial directors. Major thanks also goes to another good friend and fiberglass fanatic, Ray Quinlan of Champaign, Illinois. And how can I forget my big-block buddies, Guy Landis and Tom Biltcliff, both of Kutztown, Pennsylvania. Pop another Yuengling for me, will ya?

Former National Corvette Museum director Dan Gale deserves my gratitude as well, as does Jim Kelsey at Bill and Dean Carlson's Klassix Auto Museum in Daytona Beach, Florida. The same goes for Jana McCoy, of Chevrolet Public Relations in Atlanta. It was Jana who has loaned me three new Corvettes over the last few years, all three making appearances on these pages. Additional support and all-around great company came from Roger and Dave Judski, of Roger's Corvette Center in Maitland, Florida, and Brent Ferguson of the Classic Corvettes of Orlando Club in Florida. Yes, Brent, that is your fuelie engine on page 12.

Also, I must throw in a plug for my younger brothers, Dave, Jim and Ken, and my brother-in-law, Frank Young, all in central Illinois. It has been their company during countless Midwest photo junkets that has helped make my long excursions thoroughly enjoyable—and far less exhausting.

Finally comes mention for all the people who really made this book possible—the various collectors who took the time to allow me to photograph their cars. And while they were at it they also made me feel very much at home. In general order of appearance they are:

1967 L88 and 1969 ZL1 Corvettes, Roger and Dave Judski, Roger's Corvette Center, Maitland, Florida; 1968 427 Sting Ray, Guy Landis, Kutztown, Pennsylvania; 1971 454 Stingray convertible, Tom Biltcliff, Kutztown, Pennsylvania; 1990 ZR-1, Ed and Diann Kuziel, Tampa, Florida; 1953 Corvette, Chip Miller, Carlisle, Pennsylvania; 1955 Corvette, Elmer and Dean Puckett, Elgin, Illinois; 1956 "Betty Skelton racer," 1956 SR-2 and 1963 Grand Sport, Bill and Betty Tower, Plant City, Florida; 1957 "Airbox" Corvette, Milton Robson, Gainesville, Georgia; 1961 "Big Tank" Corvette, Elmer and Sharon Lash, Champaign, Illinois; 1963 Z06 Sting Ray, Bob Lojewski, Cook County, Illinois; 1965 fuel-injected Sting Ray, Gary and Carol Licko, Miami, Florida; 1965 396 Sting Ray, Lukason and Son Collection, Florida; 1967 L71 Sting Ray convertible, Chet and Deb Miltenberger, Winter Park, Florida; 1968 L89 Sting Ray, Elmer and Dean Puckett, Elgin, Illinois; 1972 LT-1 Stingray, Steve and Nora Gussack, Winter Springs, Florida; 1978 Limited Edition Indy Pace Car replica, George and Judi Augustine, New Smyrna Beach, Florida; 1984 serial number 00001 Corvette, Dick Gonyer, Bowling Green, Ohio; 1988 35th anniversary Corvette, Don and Denise Sanzera, Marco Island, Florida; 1991 Callaway Twin Turbo Speedster, Milton Robson, Gainesville, Georgia; 1993 40th anniversary ZR-1 "Black Widow," Jerry Crews, Longwood, Florida.

A hearty thank you to everyone.

INTRODUCTION

FOUR DECADES OF LIFE IN THE FAST LANE

Hard to believe, isn't it? Chevrolet's Corvette, that fantastic plastic flight of fancy, is now in its forties, an age when more than a few of us mere mortals are already rapidly rolling towards the bottom of the hill. Not so for this country's only true sports car. Middle-aged status notwithstanding, Chevy's famed fiberglass two-seater is still out there proving it all night, something this sexy plaything has been doing with exceptional flair longer than any other American performance machine. Class? Convenience? A kick-ass reputation as both a straight-line screamer and curve-hugging road rocket? It's all in there in generous portions no other sports car can match. At least not for the price.

Isn't it amazing what a difference four decades can make? Forty years of rich Corvette heritage was marked in 1993 by a run of exclusive anniversary models, all painted Ruby Red Metallic. In 1953, Chevrolet's little fiberglass two-seater was a roadster in the true sense of the word—side curtains instead of windows and no outside door handles. Today, the Corvette offers as much comfort and convenience as it does performance.

That's not to say competitors haven't tried over the years. Rivals to the throne have come and gone, some weakly, others with bravado, but all in short-lived fashion. Muntz Jet. Nash-Healey. Kaiser-Darrin. Dual-Ghia. A.C. Cobra. And to a degree, Shelby's Mustang and American Motors' AMX. No, Dodge fans, today's Viper doesn't even come close to earning equal billing. Brutish it is. A proven, complete package it's not. Come back in a year or two (or 10), and then maybe we'll talk.

Let's not forget the foreign challengers. M.G. Austin-Healey. Jaguar's various XKs. Porsche. Aston-Martin. Tough competition all, but none of Europe's best (or Japan's) have ever been able to touch the Corvette in the category near and dear to the hearts of most Yankee buyers—value. Look up the dictionary listing for "most bang for the buck" (in world-class sports car terms) and you'll likely see a picture of a Corvette. If

Fuel injection appeared as a Corvette option, at a cost of $484.20, for the first time in 1957. Various modifications raised "fuelie" output over the years before the Rochester injection equipment was deleted after 1965.

you take $40,000 out today and come home with as much relatively roomy (again, in sports car terms) comfort, prestigious pizzazz and pulsating performance as offered by Chevy's all-American two-seater, you've undoubtedly had to settle for something less than new. Or far less exciting.

And to think General Motors' decision-makers almost gave up on Chevrolet's fiberglass fantasy for lack of interest after about two years on the road.

When the first Corvette rolled off its makeshift assembly line in Flint, Michigan, on June 30, 1953, it certainly fit the classic sports car mold, what with its somewhat crude folding top and plastic side curtains in place of roll-up windows. All 300 first-edition roadsters were painted Polo White with red interiors. And all were powered by a muscled-up version of Chevrolet's yeoman "Stovebolt" six-cylinder backed by a less-than-desirable two-speed Powerglide automatic.

Obviously, "limited" was a fair description in more ways than one. And adding exterior paint choices in 1954 did little to change that fact. With nearly a third of the 3,640 1954 Corvettes built sitting unsold at year's end, many at Chevrolet were wondering if it wasn't just best to cut and run.

But they didn't, and the Corvette quickly found its niche after V-8 power and a manual transmission were added to the mix in

That large, gold "V" tacked over the Chevrolet emblem on this 1955 Corvette signifies the presence of a 265ci V-8, the engine that basically saved America's only sports car from quick extinction.

It certainly fit the classic sports car mold. ■

1955. Nearly four decades later, Chevrolet rolled out its 1 millionth Corvette—appropriately enough, an Arctic White convertible with red interior—on July 2, 1992, proving that someone at GM knew what they were doing. Among others, names like Harley Earl, Ed Cole, Zora Arkus-Duntov, Bill Mitchell, and two Daves, McLellan and Hill, quickly come to mind.

It was long-time GM styling mogul Earl who first campaigned for the little two-seat sportster in the early 1950s. Cole supplied all-important support for the project early on, first as Chevrolet's chief engineer, then as the division's general manager beginning in July 1956. And Duntov, the so-called "father of the Corvette", needs no introduction. After Zora retired as Corvette chief engineer in 1975, Dave McClellan took his place, followed by Dave Hill in 1992. As for Mitchell, he replaced Earl as head GM stylist in 1958, then was responsible for the startling Sting Ray transformation unveiled for 1963. Of course, the list of other prime movers and major players both inside GM and out runs much longer than this. But they must remain in the shadows for now to allow the cars themselves to shine in the spotlight supplied by this ninety-six-page mini-epic.

A similar challenge appeared when faced with trying to properly honor the major milestones in the Corvette's rich forty-odd-year history. Although beginning the story is easy enough, where do you end? Perhaps discounting, with all gentleness, the comparatively weak-kneed models built in the performance-starved late 1970s and early 1980s, hasn't damn near every Corvette built been a truly great American car? And in the years since—especially after Chevrolet introduced the wonderfully efficient, exceptionally potent second-generation LT1 small-block V-8 in 1992—hasn't each succeeding model briefly represented the best Corvette yet?

Z06

As fuel injection was falling by the wayside in 1965, the Corvette was receiving its first big-block V-8, the 396ci Mk IV. Chevy's "Mk motors" had debuted in 427ci racing form at the Daytona 500 in 1963. As a Corvette option, the 396 Mk IV rated at a healthy 425hp, enough muscle to make the Sting Ray America's most powerful production car this side of Shelby's 427 Cobra Euro-Yankee hybrid.

LEFT
"If you take that off," claimed GM styling chief Bill Mitchell, "you might as well forget the whole thing." Mitchell was referring to his pet Sting Ray styling element, the so-called "stinger" that separated the all-new 1963 Corvette's rear windows. Among others, Zora Duntov didn't like the split-window theme due to the way it hindered rearward vision. Despite Mitchell's emotional appeal, Chevrolet did "take that off" for 1964. And didn't forget the whole thing.

Nonetheless, many great moments do stand out more than others. And although determining which of these ranks among the greatest may represent a tough task, following the everyday Corvette legacy through its various ups and downs and ins and outs, isn't all that difficult.

The Corvette was built in St. Louis, Missouri, from December 1953 through July 1981, and in Bowling Green, Kentucky, from June 1981 to present. It has evolved through four definable "generations," beginning with the solid-axle variants of 1953 through 1962. The so-called "mid-year" models of 1963-67 make up the second generation, followed by the third running from 1968 to 1982. After the "hiccup" year of 1983, an all-new Cor-vette appeared for 1984 to kick off the fourth generation, which is coming to an end as we speak. Chevrolet's "C-5"—"5" for fifth generation—awaits its debut, either as a 1997 or 1998 model, depending on how soon Dave Hill's men decide to stop teasing us.

CORVETTE

A new Corvette body for 1968 (right) drew both raves and complaints for its sexy shape, some saying it was "painfully too American." Truly American in this case are the two big-block V-8s beneath these Corvettes' hoods. The silver 1968 has the 427 derivative of Chevy's Mk IV big-block; the blue 1971 convertible has the larger 454, introduced for 1970. Big-block Corvette production lasted until 1974, when the last 454 Sting Ray was built.

All Corvettes were convertibles through 1962, with an optional removable hardtop available after 1956. Roll-up windows were added that year as well, making Chevrolet's sports car more socially acceptable from a Yankee perspective. Independent rear suspension and a coupe model first appeared in 1963, when all Corvettes also became Sting Rays—a name that stuck until it fell by the wayside after 1976. From 1969 to 1976, "Stingray" was used. Beginning in 1976, all Corvettes were coupes as the droptop model was deleted, only to reappear in 1986. Single headlamps were used from 1953 through 1957, followed by dual units in 1958. Hideaway headlights have been the norm since the Sting Ray's appearance in 1963.

LEFT
A potent pair indeed. Both the aluminum-head L88 (back) and all-aluminum ZL1 427s were jokingly rated at only 430hp. Actual output zoomed past 500 for these race-ready rockets. Only twenty 1967 L88 coupes were built, while two 1969 ZL1s escaped Chevrolet Engineering.

Easily recognized by its widened tail section, added to help house some serious rubber in back, Chevrolet's first ZR-1 Corvette emerged in 1990 to do battle with the world's best sports car. With a top end of nearly 180mph, the 375hp LT5-powered ZR-1 was an able competitor. LT5 output jumped to 405hp in 1993. ZR-1 production ended two years later.

As for power, Chevrolet's ground-breaking 265ci overhead-valve V-8 replaced the valiant, yet disappointing six in 1955. Optional dual four-barrel carbs appeared the following year. Then came fuel injection in 1957, the same year the enlarged 283 small-block V-8 and available four-speed manual transmission debuted. Standard displacement grew to 327ci in 1962, when only a single four-barrel was offered atop the three available carbureted V-8s. The last "fuelie" Sting Ray was built in 1965 as the Corvette's first big-block V-8 was appearing as the top power option. Originally offered in 396ci form, the legendary Mk IV big-block bully was bumped up to 427ci in 1966, then to a whopping 454ci in 1970.

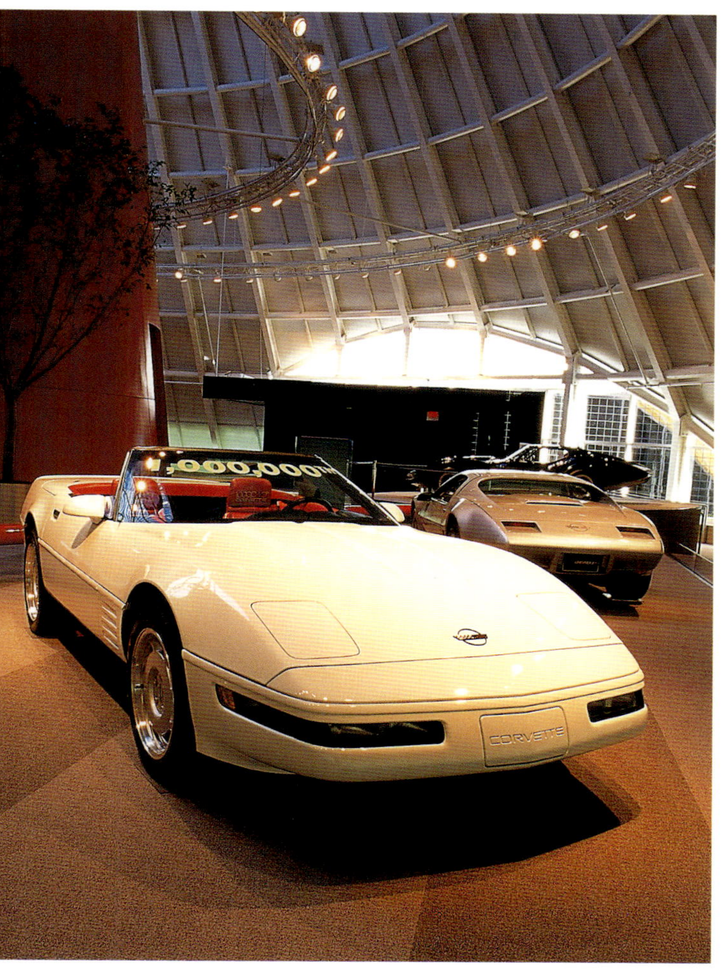

On July 2, 1992, the Bowling Green assembly plant in Kentucky rolled out the celebrated 1 millionth Corvette, a white convertible with red interior. Chevrolet later donated that car to Bowling Green's National Corvette Museum, where it resides today.

Independent rear suspension and a coupe model first appeared in 1963, when all Corvettes also became Sting Rays, a name that stuck until it fell by the wayside after 1976. ■

Meanwhile, small-block performance was progressing at full speed as the 370hp LT-1 350 was introduced for the 1970 Corvette. The most powerful carbureted small-block (the fuel-injected 327 rated at 375 horses in top tune and the ZR-1's injected LT5 later hit 405hp) ever offered under that long fiberglass hood was the first-generation LT-1. It lasted but three years before falling victim to rising insurance costs, ever-tightening emissions standards and tougher federal safety specifications. The dreaded fuel crunch of the 1970s also helped bring down the 454 Stingray, with the last big-block Corvette rolling out of St. Louis in 1974.

Carburetors disappeared after 1981 as "Cross Fire" injection was made standard for the 1982 Corvette, a car which, like its early forerunners, came only with an automatic transmission. After a one-year drought, a four-speed manual returned to its proper place on the options list when the redesigned fourth-generation Corvette appeared as a 1984 model. A true "fuelie" Corvette returned in 1985 with the arrival of Bosch's tuned-port injection setup, which carried over atop the sensational 300hp, 5.7-liter, second-generation LT1 V-8, introduced in 1992. The TPI LT1 remained the Corvette's heart and soul until replaced by the sequential-port fuel-injected LT1 in 1994.

Improvements in 1994 made the LT1 feel even stronger even though advertised output remained at 300 horses. Modifications included trading the previously used multi-port fuel injection for a sequential-port setup. Speed-density fuel calibration was also superseded by a more precise mass-air flow calibration system.

RIGHT
Beneath the 1993 40th anniversary Corvette's Ruby Red Metallic skin beat the heart of the exceptional 5.7-liter LT1 small-block V-8, a 300hp thriller first introduced in 1992. By 1994, the LT1 had become the backbone of Chevrolet's performance lineup, powering everything from Z28 Camaro to Impala SS.

Hasn't damn near every Corvette built been a truly great American car? ■

Honors in recent decades include three trips around the legendary Brickyard in Indianapolis as the prestigious pace car for the Indy 500. In all three years, 1978, 1986, and 1995, street-going replicas of those official Indy pacers were sold to the public. Special high-profile models were also created to mark the Corvette's 25th, 35th and 40th birthdays in 1978, 1988 and 1993, respectively. And equally special "Collector Editions" were offered in 1982 and presently for 1996. New also for 1996 is the Corvette Grand Sport, a modern-day, regular-production commemoration of its all-out race-ready namesake of 1963. Power for the new Grand Sport comes from a 330hp LT4 V-8.

But easily the greatest moment in Corvette history came some six years ago when the feared and revered ZR-1 debuted as a 1990 model. The ZR-1's high-tech aluminum LT5 V-8, a 375hp 5.7-liter small-block with dual overhead cams and 11:1 compression, quickly helped remind many by-standers of some earlier exotic Corvettes. Prime examples include the 425hp LS-6 of 1971, the all-aluminum ZL-1 of 1969 and the aluminum-head L88s built from 1967 to 1969. But unlike those grumpy, race-bred, ultra-low-production beasts, the ZR-1 was an animal that was easy enough to get along with on the street, yet more than capable of ripping the lungs out of any and all stoplight challengers. And it also stuck around long enough to make a noticeable impression, although market pressures finally helped bring about the ZR-1's demise in 1995.

Even without the ZR-1 to lead the way, Chevrolet's Corvette remains in 1996 every bit as great as so many of its legendary ancestors. Four decades down the road, and America's only sports car is still rockin' as strong as ever. How many of us today wish we could say that?

RIGHT
Forty-three years after its birth, America's only sports car is still running strong for 1996. Big news this year involves the debut of Chevrolet's optional 330hp LT4. Standard power still comes from the tried-and-true 300hp LT1.

HUMBLE BEGINNINGS

Okay, so it wasn't all that hot of a car, what with its somewhat ho-hum six-cylinder powerplant, decidedly non-sporty two-speed Powerglide automatic transmission and yeoman chassis. But if you want to talk about the greatest of the great from the Corvette bloodline, you have to kick off the discussion somewhere.

Seriously, though, Chevrolet's first Corvette certainly deserves more than a backhanded compliment or two. Its ground-breaking status as this country's first true "mainstream" sports car built by a major manufacturer is enough to forever honor those 300 Polo White 1953 roadsters as milestones in American automotive history.

All 300 1953 Corvettes were painted Polo White with red interiors. And like so many of its sports car rivals from Europe, Chevy's fiberglass two-seat roadster didn't have exterior door handles or roll-up windows. Aftermarket companies would soon be offering an available bolt-on hardtop, something Chevrolet would make official in 1956; the same year roll-up windows and external door handles were added to the mix.

That the major manufacturer was Chevrolet—a leader in frugal practicality—only served to help make the Corvette's birth even more historic. After 1953, more and more buyers changed the way they looked at that signature Bow-Tie. This was due to the debut of Chevy's little two-seat sportster, followed two years later by the arrival of the division's first modern overhead-valve V-8. And from there, the low-priced market would never be the same.

Before 1953, the low-priced field was the last place a speed-conscious customer looked for performance, let alone sports car performance. Yet at the time, sports cars were reasonably popular in America, most of them being imported from Europe. All previous attempts to market Yankee reactions to this European postwar invasion had basically consisted of low-production independent efforts or special hybrids made up of American engines in foreign bodies. Chevy's entry into this field, however, was red, white and blue through and through.

Much work went into designing the first Corvette's dual exhaust outlets to prevent staining the bodywork in back. The car's folding top didn't do much for those low lines, credited originally to long-time GM styling head Harley Earl.

Although much of the credit for developing the Corvette falls on the shoulders of long-time chief engineer Zora Duntov, he wasn't working for General Motors when the project began. By the time he joined Chevrolet's research and development team in May 1953, the wheels were already turning and initial production start-up was one month away. While Duntov would soon boldly make his presence known, it was actually Harley Earl, legendary GM styling head from 1927 to 1958, who may well represent the true "father of the Corvette."

Earl began toying with the idea of a regular-production sporty two-seater in the fall of 1951, after creating a pair of high-profile two-place Buick showcars. A plaster model was ready by April 1952, and support for the project came the following month from Chevrolet chief engineer Ed Cole. Chevy engineers were brought into the fray in June, kicking off a mad rush to build a showcar prototype for GM's upcoming Motorama at New York's Waldorf-Astoria hotel in January 1953.

Under Earl's direction, designer Robert McLean laid out the basic platform on a 102-inch wheelbase. The chassis was a mixture of stock Chevy parts (front suspension), specially adapted off-the-shelf components (steering and rear axle) and newly designed pieces (the rigid X-member frame). On top went a definitely unique fiberglass shell, supplied once regular production began by

Along with painful fact that all 1953-54 Corvettes used Chevrolet's mundane two-speed Powerglide automatic transmission, customers also found fault with the location of the tachometer, hidden here behind the steering wheel in the center of the dash. Most agreed the passenger had a better view of rpm readout than the driver.

the Molded Fiber Glass Body Company of Ashtabula, Ohio. Although quickly dated as the style-conscious 1950s marched on, the original Corvette image was well-received in 1953, since it was considered both modern and sporty.

For power, Cole's engineers tweaked Chevy's tried-and-true 235ci six-cylinder up to 150hp from the maximum 115 horses

Its ground-breaking status as this country's first true "mainstream" sports car built by a major manufacturer represents grounds enough to forever honor those 300 Polo White 1953 roadsters as milestones in American automotive history. ■

produced in passenger car applications. A boost in compression from 7.5:1 to 8:1, a bumpier mechanical cam, dual exhausts, and three Carter carburetors on a special side-draft aluminum manifold did the trick. As mentioned, Cole's men then backed up the "Blue Flame Six" with a floor-shifted Powerglide automatic, a move that became the largest target for slings and arrows once the exciting Corvette hit the road in the summer of 1953.

Chevrolet took a defensive stand right away concerning the Corvette's automatic-only status. According to research and development head Maurice Olley, "the use of an automatic transmission has been criticized by those who believe sports car enthusiasts want nothing but a four-speed crash shift. The answer is that the typical sports car enthusiast, like the 'average man,' is an imaginary quantity. Also, as the sports car appeals to a wider and wider section of the public, the center of gravity of this theoretical individual is shifting from the austerity of the pioneer towards the lux-

Additional exterior colors were added to the Corvette appeal in 1955, the same year Chevrolet's first overhead-valve V-8 appeared. While the six-cylinder remained available, up five horses that year, the majority of 1955 Corvettes featured the 265ci V-8. And some cars late in the year were equipped with the newly offered three-speed manual transmission.

Chrome dress-up only helped sweeten the pot for V-8 Corvette buyers in 1955. With a four-barrel "Power Pak," the 265 V-8 was rated at 195hp beneath a fiberglasss hood.

Three sidedraft carburetors and a split exhaust manifold represented the most obvious modifications that helped transform Chevy's durable "Stovebolt" six into the 1953 Corvette's Blue Flame six. Output was 150hp for the 235ci in-line powerplant.

ury of modern ideas." Concluded Olley, "there is no need to apologize for the performance of this car with its automatic transmission."

Others weren't so sure. "That statement should get a rise from 100,000 *Road & Track* readers," wrote *R&T's* John Bond, who nonetheless generally praised the 1953 Corvette for its performance from a ride and handling perspective. The automotive press, however, pointed out that the Corvette would never be able to compete with foreign rivals on a track thanks to that damned Powerglide. On the street, however, it was another story. According to *Road & Track's* test, an early six-cylinder Corvette could do 0-60mph in 11 seconds and the quarter-mile in 17.9 seconds—not bad at all for the time. Top end listed at 107mph.

Generally speaking, it was a decent start for a new car hastily created, basically from scratch, to compete in a completely unfamiliar field. Nonetheless, American buyers were not entirely impressed and Corvette sales lagged well below projections through 1954, causing GM officials to consider killing the project. Decision time came in 1955, a year when only 700 Corvettes were built. Hope for the future, however, had already arrived in the form of the Corvette's first V-8.

After nearly a third of the 3,640 Corvettes built for 1954 remained unsold at year's end, Chevrolet officials pulled in the reins, building only 700 more for 1955.

Work had begun on a V-8 Corvette—using Chevrolet's prototype for its all-new OHV V-8 scheduled for passenger car debut in 1955—in the spring of 1954, under the direction of performance development head Mauri Rose. While adding V-8 power to the package was undoubtedly the right thing to do to keep the project alive, the appearance in February 1954 of a plastic mockup from Ford featuring a concept soon to be called "personal luxury," also helped. That mockup was for the legendary Thunderbird which debuted in October 1954 with, among other attractions, a standard V-8 beneath the long, scooped hood. Although the two-seat T-bird wasn't exactly direct competition, it was more than enough of a threat to inspire rapid-fire reaction in the Chevrolet camp.

So, an optional V-8 joined the standard 155hp six in the Corvette lineup for 1955. At 265ci, Chevrolet's new OHV V-8 was rated at 195hp beneath a fiberglass hood, thanks to the addition of a four-barrel "Power Pak" and special cam. Corvettes equipped with the 265ci V-8—the vast majority in 1955 were—were identified by the large "V" emblem added to the "Chevrolet" fender script. A long-awaited, three-speed manual transmission was promised for 1955, but it didn't arrive until well into the model year. Estimates put manual transmission 1955 Corvette production at perhaps 70 or 80.

Predictably, performance improved considerably with two more cylinders to help pull the load. According to a *Road & Track* test, rest to 60mph required only 8.7 seconds for the 1955 V-8 Corvette, which topped out at 119mph. A quarter-mile went by in a relatively scant 16.5 seconds. Although most critics still complained about the brakes and ever-present Powerglide, it was clear the Corvette was on the right track to recovery, thanks to the welcomed power boost.

All other complaints would be addressed soon enough; in 1955, simply saving the car from extinction was the sole goal. And the first V-8 Corvette helped do just that.

CHAPTER 2 | 1956-62

RACING IMPROVES THE BREED

Chevrolet officials from the beginning chose to throw in a disclaimer when referring to their first-edition Corvette in print, stating that Chevy's polite two-seater was not intended for use as a "racing sports car." As it was, they really didn't need to point out what quickly became obvious to both buyers and innocent bystanders alike. Critics in the press were more than willing to explain how the wimpy Powerglide automatic would never succeed on a track. And even though it was by no means a weakling, the early Corvette's Blue Flame Six certainly had room for improvement.

A few chassis tweaks, an all-new OHV V-8 mated to a manual transmission, and the emergence of chief Corvette engineer Zora Arkus-Duntov, and America's only sports car was almost totally transformed. Following the 265ci V-8's debut in 1955, an exciting new body appeared for 1956. Appeasements to Yankee sensibilities produced roll-up windows, external door handles and an optional removable hardtop.

More importantly, at least from a horsepower hound's perspective, the 265ci V-8 was boosted to 225hp with the addition of RPO (regular production option) 469, made up of two Carter four-barrel carburetors. And if you really wanted to get serious, you could order the so-called "Duntov cam," a high-lift bumpstick specified "for racing purposes only." Clearly, Chevrolet's attitude about the Corvette had changed. Listed under RPO 449 or 448, depending on your source, the Duntov cam was only available with RPO 469. No official advertised horsepower figure was given, although most sources put output for the special-cam engine at 240hp.

Featuring unique aluminum trim throughout to help save weight, this Smokey Yunick-prepped 1956 Corvette was one of the three cars taken to Daytona in February 1956 for NASCAR's annual Speed Week trials. Originally driven on the sands by Betty Skelton, the car today resides in Bill Tower's noted Corvette collection. Tower also owns an SR-2 and 1963 Grand Sport.

Various modification tricks used on the "Betty Skelton racer" included these brake-cooling ducts. Feeding outside air through vented backing plates into the drums, this setup would soon become a regular Corvette option. These ducts would later come to be called "elephant ears."

RIGHT
Bill Tower's 1956 SR-2 was the second of three built. After Harley Earl ordered the first SR-2 for his son Jerry, Bill Mitchell requested one be built for himself. SR-2 number two's competition debut came in 1957 at Daytona, where Buck Baker recorded a flying-mile time of 152mph.

Firestone

Along with the optional removable hardtop, this 1957 fuel-injected Corvette also features wider 15x5.5 wheels, as identified by the small hub caps. This fuelie is an "Airbox" car as well, meaning it is equipped with special ductwork included to help supply cooler, denser outside air direct access to the Rochester injection setup. Only 43 Airbox Corvettes were built for 1957.

Now armed with a race-ready rocket, Duntov set out to prove that his baby was no longer intended solely for streetside duty. In February 1956, he took a three-car team to Daytona Beach, Florida, for NASCAR's annual Speed Week trials. Joining him were veteran race driver John Fitch and champion aerobatic pilot Betty Skelton. By the time the three finally slowed down, Duntov had established a new sports car flying-mile standard of 150.533mph, while Fitch had set a two-way record at 145.543mph. Skelton managed a 137.773mph two-way clocking.

Corvettes quickly established themselves as able-bodied competitors in Sports Car Club of America (SCCA) stock-class racing. ■

Lightened throughout, the SR-2 also featured many modifications that may or may not have been retrofitted, including a four-speed manual transmission, equipment that didn't become a Corvette option until 1957. Notice the twin racing windscreens and the cooling louvers added to the hood.

Fitch then led a four-car effort south to Sebring for the 12-Hour endurance event in March with less impressive results. Even Duntov knew the Corvette still wasn't ready to take on Europe's best in endurance competition. Going flat-out in speed trials was one thing; slowing and speeding up around twists and turns was an entirely different ball game which Europeans knew how to play all too well. From the beginning, Duntov recognized he would need much more of a sports car to beat foreign rivals on their turf.

Nonetheless, Corvettes quickly established themselves as able-bodied competitors in Sports Car Club of America (SCCA) stock-class racing. Armed with an ever-growing arsenal of hot factory parts such as special brakes and beefed suspension, Dr. Dick Thompson first proved that Chevy's fiberglass two-seater could indeed be used as a racing sports car. In 1956 he claimed his first SCCA production-class championship at the wheel of a Corvette. Thompson won again in 1957, 1962 and 1963, and added an SCCA C-Modified title in 1960.

The Airbox arrangement's cool-air plenum can be seen at the top of this photo directly inside the fender—it's connected by that flexible duct to the Rochester unit. Listed under RPO 579E, the Airbox 283 V-8 rated the same as the top 1957 fuelie small-block, 283hp.

1
CORVETTE

Underhood crowding created by the Airbox ductwork meant the tachometer drive had to be relocated, which in turn meant the tach itself had to be moved from its typical spot in the dash to atop the steering column. The column-mounted tach is one of the easiest clues to the identity of an Airbox Corvette. Notice the medallion located in the opening left behind by the remounted tach.

Duntov's SS, designated XP-64, was hastily created beginning in July 1956 for competition at Sebring in March 1957. ■

A heavy-duty racing-type suspension and specially cooled racing brakes officially appeared as a regular-production Corvette option in 1957, as did a welcomed four-speed manual transmission. Appearing for the first time was a Positraction rear axle and wider wheels. Reportedly, some racers were also supplied with a few oversized fuel tanks, a feature that would eventually end up on the options list in 1959. But the biggest news was the arrival of Ramjet fuel injection.

LEFT
Truly innovative throughout, the magnesium bodied SS racer of 1957 nonetheless fell victim to hasty development. Even though it possessed the power to compete with Europe's best, the car carried too many gremlins with it onto the track at Sebring in March 1957. It retired after only 23 laps.

Once atop the 1957 Corvette's enlarged 283ci small-block, Ramjet fuel injection, supplied by Rochester, boosted output to 283 horses in top form. It was the second time a Detroit V-8 reached the one-horsepower-per-cubic-inch milestone—Chrysler had offered an optional 355hp 354 Hemi V-8 for its 300B luxury cruiser the year before. Although a bit finicky, the fuelie 283ci was capable of powering a 1957 Corvette from zero to 60mph in less than six seconds. Simply sizzling.

A Chevrolet team returned to Sebring in 1957, but early Corvette competition efforts weren't limited to production-class racing. The distinctive SR-2 "prototype racers" had emerged the year before. Conceived by Harley Earl for his son Jerry, the first Corvette SR-2 featured a custom fiberglass shell on a race-ready chassis. The SR-2 incorporated all the heavy-duty suspension and brake tricks Fitch had put to the test at Sebring in 1956. In Chevrolet engineering lingo, those beefed parts were known as "SR" components, which may or may not have stood for "Sports Racing" or "Sebring Racer." Either

Powering the SS was a modified fuel-injected small-block with aluminum heads and a magnesium oil pan. Additional tweaks included revised valves and a special injection manifold. Dyno tests in 1957 put output at 307hp.

Veteran driver John Fitch and co-driver Piero Taruffi sat here on race day at Sebring in 1957 after Stirling Moss and Juan Fangio both withdrew from the SS team just prior to the event. One of the SS racer's many glitches involved heat buildup inside the cockpit. Both Fitch and Taruffi were literally boiled as the magnesium shell didn't dissipate heat like its fiberglass counterpart.

"Sports Racing" or "Sebring Racer." Either way, Fitch's Corvette factory racers were the first "SRs," meaning any variation to follow was naturally the second. Thus the "SR-2."

Jerry Earl's SR-2—shop order number 90090—was built in about four weeks, leading some to believe its modified body was simply dropped right onto an existing Sebring chassis. Supposedly, an off-the-lot 1956 Corvette went into Engineering in May and came out as the SR-2 in June. Presto.

On top of that race-ready chassis was a metallic blue body wearing an extended snout and bright aluminum bodyside cove panels. Louvers on the hood and vents in each door were functional; the former cool-

Once atop the 1957 Corvette's enlarged 283ci small-block, Ramjet fuel injection, supplied by Rochester, boosted output to 283 horses in top form. It was the second time a Detroit V8 reached the one-horsepower-per-cubic-inch milestone. ■

ing the engine, the latter cooling the rear brakes. Twin short windscreens were used up front, while a small tailfin was added down the center of the decklid in back.

Racing modifications included cutout exhausts, an oversized fuel tank and Halibrand knock-off mag wheels. Interior treatment, however, was not at all like a race car's. Looking more like a showcar, Earl's SR-2 featured blue vinyl seats, custom instrumentation in a stainless-steel dash panel, and a wood-rimmed steering wheel. But all the extra flash translated into extra weight, which resulted in a disappointing competition debut at Wisconsin's Elkhart Lake in June 1956.

Earl's SR-2 was then lightened, but real success didn't come until the car was sold in 1957 to Jim Jeffords, driver for Chicago dealer Nickey Chevrolet. By that time, the first SR-2 featured a larger, reshaped tailfin offset to the driver's side to serve as a headrest and rollbar. The new design was added at Earl's request, after he saw it on the second SR-2 built for Bill Mitchell. As for SR-2 number one, in Jeffords' able hands it raced to a SCCA B/Production championship in 1958.

Mitchell's red-and-white SR-2 debuted in February 1957 during Daytona's Speed Week trials. There, Buck Baker recorded a flying-mile speed of 152.886 mph in the "high-finned" car, which finished sixteenth at Sebring a month later.

A third "low-finned" SR-2 was built for GM president Harlow Curtice. Basically a stock 1956 Corvette underneath, Curtice's SR-2 was meant only for the show circuit. Metallic blue paint was again used, as were Dayton wire wheels and a removable stainless steel hardtop.

All three SR-2s still survive, after passing through various owners' hands and experiencing a modification or two along the way. Originally powered by dual-carb small-blocks, the trio are now fuel injected. Four-speed transmissions are also on board even though, like Rochester fuel injection, a four-speed didn't appear as a Corvette option until 1957. Some bystanders believe

This is one of Larry Shinoda's drawings for the CERV 1 test vehicle. Built in 1960, CERV 1, among other things, established many of the design parameters for the 1963 Sting Ray's independent rear suspension.

RIGHT
Penned by stylist Larry Shinoda, the XP-755 Mako Shark was built in 1961 as a personal car for Shinoda's boss, Bill Mitchell. XP-755 borrowed many of its lines from the XP-720 project, which was the prototype for the all-new 1963 Sting Ray. When another Mako Shark styling exercise appeared in 1965, Mitchell's car became known as the Mako Shark I. Today, the Mako Shark I resides at the National Corvette Museum in Bowling Green, Kentucky.

the prototype installation tale, while others simply believe the fuelie/four-speed equipment came via a retrofit.

While SR-2 racers represented independent efforts, the next great Corvette competition project was a factory job from top to bottom. As previously mentioned, Duntov always wanted a world-class racing machine. And his first attempt to transform his ideal into reality came in the form of the SS racer of 1957.

Duntov's SS, designated XP-64, was hastily created beginning in July 1956 for competition at Sebring in March 1957. It

Adding the oversized 24-gallon fuel tank meant the removable hardtop option was also required since the tank took up the space behind the seat where the folding top normally resided. The oversized tank was first officially offered for 1959, although some reportedly had been supplied to racers as early as 1957.

was an all-out, purpose-built machine, with its low-slung tubular space frame and lightweight magnesium body. Aluminum was used throughout, including the gearbox, radiator and cylinder heads on the modified 283 fuelie V-8. Coilover shocks were at all four corners, an independent de Dion rearend was in back, and the brakes were finned drums with the rear pair mounted inboard on the differential to reduce unsprung weight. An innovative, servo-controlled booster system was also incorporated to help prevent rear wheel brake lockup. Along with the one beautiful blue SS built in Chevy Engineering, Duntov's crew also fashioned a crude fiberglass-bodied test mule counterpart.

LEFT
This 1961 fuelie features various racing-inspired options, including wide wheels, quicker steering and heavy-duty suspension with specially cooled brakes. But rarest of the bunch is LPO (limited-production option) 1625, the so-called "big tank." Notice the exposed fuel filler cap behind the door—it represented one of the modifications made to mount the larger tank behind this 1961 Corvette's seat.

Chevy's vaunted small-block displaced 265 cubes when introduced in 1955, then grew to 283ci in 1957. In 1962, the last year for the solid-axle first-generation Corvette, the small-block again grew, this time to 327ci. This is the top carbureted 327 for 1962. Output was 340 horses, compared to 360 for its fuel-injected counterpart. Only single four-barrel carburetors were offered in 1962; dual carbs were last offered in 1961.

On paper, the SS project certainly looked promising. And early tests of the white SS mule at Sebring were impressive. But too much work was rushed trying to make the tight Sebring deadline. Problems appeared almost immediately on race day for the blue SS, including rising cockpit temperatures, fading brakes and a loose rear suspension. Only twenty-three laps were completed before the SS was forced to retire from the 12-Hour endurance event.

A few chassis tweaks, an all-new OHV V8 mated to a manual trans, and the emergence of chief Corvette engineer Zora Arkus-Duntov later and America's only sports car was nearly totally transformed. ■

Then, any hopes of going back to the drawing board and trying again at Le Mans later that summer were dashed, once the Automobile Manufacturers Association (AMA) issued its so-called ban on factory racing involvement. One can only wonder what might have been, especially after Duntov's SS hit 183mph at GM's Phoenix proving grounds in December 1958.

The lone blue SS made one more cameo appearance in February 1959, turning a 155-mph lap around NASCAR's brand-new Daytona International Speedway during opening ceremonies. Later, the car was donated to the Indianapolis Motor Speedway Hall of Fame Museum where it still resides today.

As for the fiberglass SS mule, it also survives, although not in its original form. During the winter of 1958-59, it was acquired by GM styling head Bill Mitchell. Its chassis was used as a base for his XP-87 Stingray racer, a car that provided more than one styling trick later used on the regular-production Sting Ray in 1963. With Dr. Dick Thompson at the wheel, Mitchell's Stingray roared to an SCCA C/Modified championship in 1960.

With the AMA "ban" in place by the summer of 1957, Chevrolet was forced to cut back on its factory racing efforts. While factory support of certain Corvette racers—namely Dr. Thompson—continued, it was primarily the covert, "back-door" type. Clearly obvious, however, was the long list of factory options still present and accounted for after 1957: big metallic brakes with special cooling ducts, quicker steering gear, bigger wheels, and beefed suspension components. Duntov even tested a set of weight-saving aluminum fuelie cylinder heads in 1960, although that option was quickly discontinued when production defects couldn't be cured. All this purposeful equipment and more was on the option list, as were loads of fuel-injected power—up to 360 horses worth by 1962.

And American sports car devotees who thought it just couldn't get any better needed only to wait another year.

MECHANIC
Bill Scott
2
AIR JACKS
OIL
Firestone

ENTER THE STING RAY

It may easily rank among the most startling transformations in American automotive history. When Zora Duntov, Bill Mitchell and crew took what was already one of this country's most startling automobiles and redesigned it for 1963, they succeeded in creating a true modern classic. It is this car that most often comes to mind when even the most casual observer thinks "Corvette."

From nose to tail, the all-new 1963 Sting Ray was a stunner. "This is the one we've been waiting for," wrote *Motor Trend's* Roger Huntington. "This is a modern sports car." Even Duntov himself was finally satisfied: "For the first time I now have a Corvette I can be proud to drive in Europe."

Zora Duntov tried again to build a world-class racing Corvette late in 1962. But plans to build 125 lightweight Grand Sports failed when GM's anti-performance overlords shot things down. Only five Grand Sports were built before the ax fell early in 1963. Two of these were later converted into roadsters.

Lower, thinner, shorter and riding on a compact 98-inch wheelbase (down from the previous 102-inch chassis), the sleek, sexy Sting Ray was nonetheless more comfortable inside than its solid-axle forerunner, thanks to repositioned seating in a redesigned frame. And innovative (at least in Yankee terms), independent rear suspension improved both ride and handling.

On top of it all was Mitchell's alarming new Corvette shell—available for the first time in coupe form—with hideaway headlights in front, and crisp, curvaceous lines throughout. It featured a tapered roofline with Mitchell's pet split-window "stinger" theme in back. Any way you looked at it, the car was a killer, although many critics—including Duntov—didn't think much of the split rear windows. So, despite Mitchell's adamant defenses, the stinger was deleted in 1964.

Initial demand for the attractive Sting Ray was so great, that adding an extra shift at the St. Louis plant couldn't even help keep up. Overall, 1963 sales (coupes and convert-

Beefy brakes were part of the Z06 deal. Included was a dual-circuit power booster; enlarged, finned drums; sintered cerametallix linings; and special cooling gear. Rubberized "elephant ear" ducts directed airflow through vented backing plates where an internal fan helped stir things around. The drums themselves were also vented. One drawback to these brutish binders was that they barely worked at all before they were warmed up.

TOP RIGHT
Grand Sport racers were amazingly stock looking inside, although a closer inspection would reveal the presence of a 200mph speedometer. Also barely noticeable just above the passenger seat is a movie camera used by Chevrolet engineers to document on-track testing action.

BOTTOM RIGHT
Beneath all that plumbing is a 377ci aluminum small-block. Those are four 58mm Weber side-draft two-barrels on a cross-ram intake. The Webers on the left feed the cylinder bank on the right and vice versa. Original plans to equip the Grand Sport with the 377 small-block were canceled when the project itself was shut down by GM's front office in January 1963. Grand Sports were first powered by less exotic fuel-injected 327s. Various other power sources followed over the years.

ibles) soared by 50 percent to a new high of 21,513 cars. Still a polite tourer in base form with a 250hp 327, backed by a three-speed manual, the 1963 Sting Ray could typically be transformed into a street killer, thanks to a long list of options. Most prominent, from a power perspective, was the 360hp fuel-injected small-block, RPO L84.

Last, but certainly not least on that list was RPO Z06, the Special Performance Equipment group. Any questions about the intentions of this potent package were quickly answered once the first six Z06 Sting Rays were delivered directly into the hands of prominent racers in October 1962. The competition debut for this race-ready Corvette came October 13 in Riverside, California, at the same event that showcased the first of Carroll Shelby's little Ford-powered Cobras. Having had their way in SCCA competition throughout the late 1950s and early 1960s, competition Corvettes were soon left in the dust by Shelby's Cobras. But how could Chevy engineers have foreseen this occurrence in 1962, especially considering the tools they were about to use?

RPO Z06 consisted of every hot part on the Corvette shelf. Mandatory features included the L84 fuelie 327, backed by its close-ratio Muncie four-speed and Positraction rearend. Initially listed were heavy-duty suspension parts, special "cerametallix" power brakes with unique cooling features, an oversized 36.5-gallon fiberglass fuel tank, and five cast-aluminum knock-off wheels. In December, Chevrolet temporarily canceled the knock-off option, due to production difficulties, and removed the big

Z06

The development of larger four-barrel carburetors, far less finicky in practice than fuel-injection, helped spell the end for the Rochester fuelie setup in 1965. The debut of the 396 big-block also contributed to fuel injection's demise. Only 771 L84 F.I. Sting Rays were built for 1965.

LEFT
While a host of racing options had been available to Corvette buyers since 1957, Chevrolet really got the ball rolling in 1963, grouping all the hottest parts together in one package, labeled RPO Z06. Top fuelie power, heavy-duty suspension and special brakes were all included. Only 199 Z06 Sting Rays were built for 1963.

A performance powerplant from head to toe, the L78 396 pumped out 425 rompin', stompin' horses. Mandatory options included the close-ratio M20 Muncie four-speed, transistorized ignition (RPO K66) and a Positraction differential. Notice the optional power brakes with dual-circuit master cylinder.

LEFT
Much less costlier than the fuel injection, and considerably more powerful, was the 1965 396 Sting Ray, the first big-block Corvette. Price for the L78 big-block option was $292.70. Production of 396 Corvettes was 2,157.

427

A distinctive triangular air cleaner was part of the L71 package. Under normal operation, the 435hp 427 was fed by the middle Holley two-barrel. Putting the pedal to the metal brought the other two carbs into play via a vacuum signal. When really wailing, the 3x2 setup sucked in some 1000cfm worth of fuel/air.

tank from the package to help whittle down RPO Z06's original $1,818.45 asking price. The 36.5-gallon tank, RPO N03, remained a separate option, able to be added to any Sting Ray coupe, Z06 or other model.

Reportedly, only 199 buyers chose a Z06 Sting Ray coupe in 1963. Amazingly, some of these customers bought these rough-and-ready rockets with intentions of driving them in everyday situations—not a smart move at all, considering the car's cantankerous demeanor. In the words of engineer/author Paul Van Valkenburgh, the Z06 Corvette "was not a car to drive on the street. It was noisy, rode hard, [and] the brakes were terrible—[they] wouldn't work at all until you warmed them up very, very well."

RPO Z06 made a brief repeat performance early in 1964, at least on paper. Ini-

LEFT
After big-block displacement was increased to 427ci in 1966, the Corvette's top performance V-8 was bumped up to 435hp in 1967, thanks to the addition of three Holley two-barrel carburetors. Listed under RPO L71, the tri-carb 427 found 3,754 buyers in 1967.

Meant only for off-road use, the L88 Sting Ray appeared in the spring of 1967 with nary a clue as to its ferocious demeanor. Price for the track-ready option was $947.90. Only 20 were built in 1967.

tially listed as a 1964 Sting Ray option, the Z06 package was quickly discontinued before any were sold. Chevrolet opted instead to offer the various Z06 components as separate options.

As brutish as the Z06 Sting Ray was, it was not the meanest Corvette built for 1963. That honor went to the Grand Sport, a purpose-built race car reminiscent of Duntov's ill-fated SS of 1957. Even though the 1957 AMA "ban" on factory racing supposedly closed the door on such shenanigans, Duntov never stopped thinking about building a world-class competition Corvette. And once performance-conscious Semon E. "Bunkie" Knudsen came over from Pontiac to be Chevrolet general manager in November 1961, Duntov was allowed just enough leeway to kick off another competition-minded project.

That project began in the summer of 1962, as Duntov's engineers began fashioning a special lightweight Corvette, based on a tubular-steel, ladder-type frame. Weight was also saved by using various aluminum components, Halibrand magnesium knock-off wheels, and a special handmade fiberglass body with super-thin panels. Some external dimensions were also changed to help improve on the stock Sting Ray's aerodynamics. Plexiglass windows, a 36.5-gallon fuel tank and enlarged wheelhouse openings for larger tires were incorporated. Brakes were large 11.75-inch Girling discs at all four corners.

Initial specifications called for a 377ci small-block fed by four Weber carbs for the Grand Sport. Early plans also mentioned a production run of 125 Grand Sports. Neither became reality. The first Grand Sport was fitted with an aluminum 327 fuelie while awaiting the 377 V-8 still in development. And before that work could be finished, GM announced a ban of its own, instructing all divisions in January 1963 to cease racing projects immediately.

Only the first five of the 125 planned Grand Sports escaped Chevrolet Engineering before that order came down. From there, each went through a steady progression of so-called independent race teams. They also underwent various mechanical and exterior modifications, taking on a varied succession of scoops, flares and engines. Both small-blocks and big-blocks were used over the years, and two of the coupes were later converted into roadsters for competition at Daytona in February 1964.

GM officials, however, stepped in again before those two roadsters could reach Florida. Clearly, Chevrolet Engineering was still supporting much of the Grand Sport racing effort, despite orders to the contrary a year before. This time GM executives instructed Knudsen to end these activities or risk his annual bonus. The five Grand Sports were then sold off. And the last major competition appearance came in 1966. Like the SS and SR-2s before them, all five Grand Sports survive today in collectors' hands.

Back on the street, the next great moment in Corvette history came in 1965, when four-wheel disc brakes were made standard equipment and two great powerplants crossed paths. Chevrolet's fuel-injected small-block—a major part of the Corvette mystique and the top power source since 1957—was offered for the last time in 1965. Its high $538 asking price was no longer justifiable, since less expensive, less finicky carbureted 327s had become nearly as powerful. Even more powerful was the all-new 396ci Mk IV V-8, the Corvette's first big-block.

Introduced early in 1965, the 396 featured

Beneath the 1967 L88's functional hood was the star of the show, the aluminum-head 427. The odd-looking air cleaner fit into special ductwork in the hood's underside from which it drew denser air from the base of the windshield. Also notice the black road draft tube running from the driver's side valve cover to behind the master cylinder. An L88 couldn't pass emissions tests since it simply vented its crankcase directly into the atmosphere.

a bullet-proof block and free-breathing cylinder heads with ball-stud rockers and large, canted valves. It was the perceived haphazard fashion in which these valves protruded upward that inspired the nickname "porcupine heads." These exceptional heads, in concert with a high-lift mechanical cam, 11:1 compression, transistorized ignition, a big Holley four-barrel on an aluminum intake, and header-type cast-iron exhausts, helped the Mk IV big-block produce 425hp.

Mounting the 425hp 396 between fiberglass fenders required various modifications, the most noticeable being a bulging hood with functional louvers. Both chassis and driveline were beefed, cooling was improved, and a close-ratio four-speed and Positraction rearend were mandatory. What did all this heavy-duty equipment add up to? According to *Road & Track*, a 396 Corvette could turn the quarter-mile in 14.1 seconds, "quicker than any other standard production car we've tested except the AC Cobra."

But if the automotive press thought that was hot, they had another thing coming. The following year, the Mk IV was bored out to 427 cubic inches. Then in 1967, three Holley two-barrel carburetors were added atop the Corvette's 427, producing 435 real horsepower.

One of the best-working progressive throttle arrangements ever seen on an American multiple-carb setup, the optional "3x2" equipment was both efficient and hot to trot. Under normal driving, the middle two-barrel worked alone, pumping roughly 300cfm of fuel/air into the lazily loping L71 big-block. With the hammer dropped, a vacuum signal brought the other two carbs into action precisely, bringing total flow to about 1000cfm as all hell broke loose.

In *Car and Driver's* words, the 3x2 setup resulted "in an astoundingly tractable engine and uncannily smooth engine response." It also resulted in some serious neck injuries. *Cars* magazine's Martyn Schorr reported a best quarter-mile run of 12.9 seconds at 111mph for the L71 Corvette. Only five seconds were required to go from 0 to 60mph. Was it any wonder *Hot Rod's* Eric Dahlquist called the tri-carb Sting Ray the "hottest 'Vette yet?"

Actually, the hottest Corvette in 1967 was another racing-inspired Sting Ray, the legendary L88. Far more cantankerous than the 1963 Z06, the 1967 L88 Corvette was clearly not meant for the street, and Chevrolet wasn't afraid to admit this fact in the least. "Because the L88 is an off-road engine," read a press release, "no provision has been made for anti-pollution control. For those states that require smog-control devices, it cannot be registered for street use in a passenger carrying vehicle." Another label found inside an L88 Sting Ray was even more foreboding. "Warning: Vehicle must operate on a fuel having a minimum of 103 research octane and 95 motor octane or engine damage may result."

While Chevrolet gave the L88 427 a token rating of "only" 430hp, actual output was probably upwards of 550 horses. Aluminum heads with large valves were part of the L88 deal, as were 12.5:1 pistons. And a huge 850cfm Holley four-barrel fed cooler, denser air by a specially ducted hood. Mandatory options included transistorized ignition, power-assisted metallic brakes, F41 sports suspension, Positraction and the indestructible M22 "Rocker Crusher" four-speed. Also included with the L88 package was RPO C48, the heater-defroster delete. Who needed such luxuries on a race track, right?

On the legendary track at Le Mans in France, one of the twenty L88 Corvettes built for 1967 impressed all with its dominating speed down the Mulsanne Straight. But a thrown rod halfway through the race ended yet another Corvette attempt at international racing glory. Not all was lost, however, as the aluminum-head L88 returned in 1968, and this time became a big SCCA winner while carrying the Owens-Corning Fiberglass banner. L88 production was 80 in 1968, followed by another 116 in 1969.

Although the second-generation Corvettes, the so-called "midyear" models of 1963-67, never produced the world-beater Duntov had hoped for, they didn't dim the original Sting Ray's reputation on the street in the least. They were great cars all.

CORVETTE
FLORIDA
LTD 91N
SEMINOLE

CHAPTER 4 1968-74

BIG-BLOCKS BOW OUT

The life span for the second-generation Corvette, the mid-year models if you will, ran five years—one more than planned. Few, if any, fiberglass fans probably cared, however, especially when presented in 1967 with what many considered to be the best of the early Sting Ray breed. And to think the last of the mid-years basically represented a stop-gap of sorts.

Work on a re-styled, thoroughly modern second-edition Sting Ray had begun early in 1965, with high hopes of making that new ideal a production-line reality for 1967. But development problems delayed that debut, forcing Duntov, Mitchell and the rest to roll out one more mid-year model before the third generation finally bowed for 1968.

Inspired by stylist Larry Shinoda's Mako Shark II showcar of 1965, the 1968 Corvette was quickly equated to a Coke bottle with its bulging front and rear quarters serving as bookends for a slimmed-down midsection. Comparing the new 1968 to its "pinched-waist" 1963-67 predecessor was akin to standing Raquel Welch up next to Lily Tomlin. Along with being some seven inches longer and considerably more shapely than its forerunner, the 1968 Corvette also scored higher in the sultry department. As *Car and Driver's* critics put it, "it's lusty, it stimulates all of the base emotion lurking deep in modern man. It is the *Barbarella* of the car maker's art"—an analogy in reference to the semi-psychedelic, surely-sexploitive 1967 Dino De Laurentiis film starring Jane Fonda.

As for the stuff of legends, top power for the lusty 1968 Corvette once more came from the 427 big-block, again available in optional 435hp L71 3x2 form. The race-ready L88s were carryovers as well, for both 1968 and 1969. And those who wanted a piece of the L88 lightweight action, without all those off-

Originally listed at 370 hp in 1970, LT-1 performance dropped to 330 horses in 1971, then was net rated at 255 hp in 1972. Along with being a true road rocket, this 1972 LT-1 is also a movie star, having appeared in the 1995 hit, *Apollo 13.*

Chevrolet's third-generation Corvette received an all-new "Coke-bottle" body for 1968. Removable roof panels were also new that year. This 1968 coupe is powered by an aluminum-head L89 427. Priced at $805.75, the L89 option was checked off 624 times in 1968.

Originally introduced for the tri-carb L71 427 in 1967, the L89 aluminum-head option not only shaved off some unwanted pounds, it also added a revised combustion chamber and different valves. Advertised output, however, remained at 435hp. RPO L89 was discontinued after the last 427 was offered in 1969.

road limitations could add the L89 aluminum head option to their L71 427. First offered in 1967, the $832.05 L89 option attracted 624 buyers in 1968, after only 20 pairs were sold the previous year. Another 390 pairs went out the door in 1969. Although valves and combustion chambers differed slightly for the lightweight L89 heads compared to their cast-iron counterparts, no change in advertised output was made when they were added to the 435hp 427.

Even more aluminum was used in 1969, when Chevrolet unleashed two ZL-1 Sting Ray coupes, the last of the truly exotic Corvettes to make it to the streets. Featuring a pair of aluminum heads atop an aluminum cylinder block, the ZL-1 427 was not for the timid—of spirit or wallet. Sticker price for RPO ZL-1 alone went well beyond four grand. Much more radical than the L88 throughout, the ZL-1 still carried the same token horsepower rating—430hp. Again, actual output soared past 500 horses, a fact quickly proven at the track where a 1969 ZL-1 Corvette wowed the press with a screaming 12.1-second quarter-mile pass. Trap speed was 116mph.

Plans for yet another high-powered aluminum-aided big-block for 1970 ran afoul of Chevrolet's latest "de-proliferation" plans. Originally listed in 1970 Corvette paperwork and tested by the press in prototype form, Chevy's stillborn LS-7 454ci big-block featured aluminum heads, 12.25:1 compres-

Identified by its nose stripe and bulging hood, this ZL1 Corvette is one of only two built for 1969, although others may have resulted from crated engines being delivered into private hands.

LEFT
Save for the air pump emission controls, the 1969 ZL1 427 looks very much like the L88. But unlike the L88, the ZL-1 has an aluminum cylinder block to accompany those aluminum heads. Chevrolet also used the all-aluminum ZL1 427 in Camaros for 1969.

RIGHT
The sexy, new 1968 Corvette measured seven inches longer than its 1967 forerunner and was nearly two inches shorter in height. Base big-block power came from a 390hp 427, as this Silverstone Silver coupe demonstrates.

Demonstrated here is the Corvette's biggest big-block, the 454ci Mk IV, which replaced the 427 in 1970. This 1971 LS5 454 Stingray is one of 5,097 built.

sion and 465hp—more than enough to put an LS-7 Sting Ray into the 13-second quarter-mile bracket, according to *Sports Car Graphic*. But growing corporate concerns over how much power was too much, led GM's front office to kill the LS-7 before it made regular production, leaving Corvette buyers to make do with the much more mundane 390hp LS-5 454.

Even with Detroit's horsepower race rounding its last turn, Chevy engineers did manage one last gasp in 1971 when they made the LS-6 454 a Corvette option. LS-6 features included open-chamber aluminum heads, a big dual-feed Holley four-barrel atop an aluminum intake, transistorized ignition, and a mechanical cam. Advertised output was 425hp. Quarter-mile performance was listed at 13.8 seconds by *Car and Driver*.

But big-block Corvettes weren't the only headline makers in the 1970s. Chevrolet's long-running small-block had been enlarged again in 1969 to 350ci. And in 1970, the 350 was the base for the Corvette's hottest small-block since the 375hp L84 327 fuelie disappeared after 1965. Initially rated at five less horses than the L84, the LT-1 350 made for a better-balanced, more road-worthy Corvette, compared to the big, bullyish, nose-heavy Mk IV models. The LT-1 lineup included a solid-lifter cam, 11:1 compression, and a Holley four-barrel on an aluminum intake.

As for performance, a 1970 LT-1 ran right along with a Porsche 911 in a *Motor Trend* test. "Off the strip and onto the road course, this is where the Porsche should reign supreme," began *MT's* report. "Here is where

RIGHT
The third-generation big-block progression began in 1968 with the 390hp L36 427 (background). In 1970, the 390hp LS5 454 superseded the 427. The following year, two 454s were offered, the 365hp 454 (shown here in foreground) and the dominating aluminum-head LS6

TURBO
454
365 HORSEPOWER

GOODYEAR
WIDE TREAD F70-15

Captured at the National Corvette Museum, this 1971 LS6 Stingray is one of only 188 built. Behind it is an 1986 "Malcolm Konner Commemorative Edition" Corvette, one of 50 created by Malcolm Konner Chevrolet in Paramus, New Jersey.

we experienced the biggest surprise of the test. The [LT-1] Corvette was just as fast, if not faster, through the corners as the Porsche."

"It's a frustrated racer," claimed *Car and Driver*, "a fact it never lets you forget." After dropping to 330hp in 1971, the last of the first-generation LT-1s appeared in 1972 net rated at 255 horses.

For small-block fans who really wanted to give Porsche drivers a run for their money, Chevrolet also offered the ZR-1 package, an LT-1 Corvette with an M-22 four-speed, heavy-duty power brakes, special suspension and an aluminum radiator. Wearing a price tag of about $1000, RPO ZR-1 was offered each year along with the LT-1. Production was only 25 in 1970, eight in 1971 and 20 in 1972. A similar equipment group, RPO ZR-2, was made available along with the LS-6 big-block in 1971. Priced at $1,747, ZR-2 equipment found its way into a mere eight 425hp Sting Rays that year.

After 1971, much of the Corvette's serious sting fell by the wayside as Detroit's era of outlandish performance came to a close. Drastically lowered compression ratios in 1971 were followed by net-rated output figures the following year. Along with the 255hp LT-1, speed-conscious customers could also pick the 270hp 454 in 1972. But the Mk IV V-8 wasn't long for the world, either. Even though the Corvette's third-generation actually ran up through 1982, an end of an era came in 1974 when Chevrolet built its last big-block Corvette.

For Corvette buyers from then on, it was a small-block or no block at all.

Air conditioning typically wasn't available along with RPO LT-1 when the option debuted in 1970. Nor in 1971. But per Zora Duntov's orders, one 1972 LT-1 was taken off the line and tested with an air conditioning installation, resulting in the availability of air-conditioned LT-1 Corvettes by the end of the year. The 1972 LT-1 shown here is that very prototype.

RIGHT
Chevrolet's first-generation LT-1 Corvette was offered between 1970 and 1972 as a nimble small-block alternative to those big-block bullies. This 1972 LT-1 is one of 1,741 built.

104
HMC
OFFICIAL PACE CAR
62nd ANNUAL INDIANAPOLIS 500 MILE RACE
MAY 28, 1978
DAYTONA
RADIAL S/R

CHEVROLET'S FIBERGLASS LEGACY LIVES ON

Let's face it, the late 1970s simply were bad years as far as Detroit performance was concerned. And America's only sports car was no exception. As if sky-high insurance costs and ever-growing safety concerns weren't enough to discourage the building of high-powered automobiles, tightening emissions standards were strangling the life out of the good ol'Yankee V-8. Once the doomed 454 big-block disappeared from the Corvette options list after 1974, drivers were left with a series of continually weakening 350 small-blocks.

But the situation started turning around by 1978, the year Chevrolet celebrated the Corvette's 25th birthday. Thanks to the optional 220hp L82 350, the 1978 Corvette was still king of the American hill, a fact not missed by *Car and Driver*. "We can happily report the 25th example of the Corvette is much improved across the board. Not only will it run faster now—the L82 version with four-speed is certainly the fastest American production car—but the general driveability and road manners are of a high order as well."

Along with that welcomed L82 shot in the arm, all 1978 Corvettes also received special 25th anniversary badges to mark the special occasion. And the 25th Corvette wore a new "fastback" rear window which both aided rearward vision and improved storage space behind the seats.

Additional commemoration was initially available through RPO B2Z, which added an exclusive two-tone silver anniversary paint scheme. Dual sport mirrors and aluminum wheels were required options, along with the paint. Production of these silver anniversary Corvettes was 15,283.

Another special 1978 model, the Limited Edition, marked that anniversary, as well as the Corvette's first appearance at Indianapolis

Chevrolet marked the first of three prestigious Indy Pace Car appearances for the Corvette with this Limited Edition model in 1978. Among a long list of features were the front air dam, aluminum wheels and rear spoiler. All 1978 Corvettes were 25th anniversary models.

It doesn't come much more distinctive than the optional silver leather interior on this 1978 Limited Edition Corvette. Not visible behind the steering wheel is an AM/FM 8-track stereo, included in the Limited Edition package.

as the prestigious pace car for the Indy 500. Priced at $13,653.21, compared to $9,351.89 for a base 1978 sport coupe, the Limited Edition Indy Pace Car replica was stuffed full of options. Included were power windows, door locks and antenna, removable glass roof panels, a rear window defogger, air conditioning, sport mirrors and a tilt-telescopic steering column. Other options included white-letter P225/60R15 tires, a heavy-duty battery, and an AM/FM 8-track stereo with dual rear speakers. A front air dam, a rear spoiler and aluminum wheels with red pinstripes completed the deal, which was originally intended to be of "limited" status, but ended up quite the contrary. When the feeding frenzy for what many felt would be a future collectible came to a close, Chevrolet sold 6,502 of these high-profile fastbacks. Today, their collector value remains minimal.

Yet another special-edition model came four years later, just as the third-generation Corvette was bowing out. Like the 1978 silver anniversary Corvette, the 1982 Collector Edition featured unique paint, this time a silver-beige finish accented by graduated grey

RIGHT
As part of a charity fund-raising effort, the National Council of Corvette Clubs raffled off the very first of the all-new 1984 Corvettes, serial number 00001. In collector Dick Gonyer's hands today, the car is still identified on the doors and windshield as it was when originally raffled. This is the only #00001 model from any Corvette generation known to survive.

CORVETTE
00001
1984
1984
00001
CHAMPION

CORVETTE
FLORIDA
HLB 553

A Corvette convertible returned in 1986 just in time to become the second fiberglass two-seater from Chevrolet to pace the Indianapolis 500. This 1986 Indy Pace Car replica resides in the Klassix Auto Museum in Daytona Beach, Florida.

decals and pinstriping. Features included "hatchback" rear glass, special emblems, exclusive "turbine" alloy wheels wearing white-letter P255/60R15 rubber, a leather-wrapped steering wheel, matching silver-beige leather upholstery, and luxury carpeting. Removable glass roof panels done in unique bronze tinting, a rear window defogger, and a power antenna were also included in the package. The price for that package? Try about $22,500, up more than $4,000 beyond the base coupe's sticker. In chief engineer Dave McLellan's words, the 1982 Collector Edition was "a unique combination of color, equipment, and innovation to produce one of the most comprehensive packages ever offered to the Corvette buyer." Production was 6,759.

LEFT
Chevrolet celebrated an important Corvette birthday for the second time in 1988, offering its 35th anniversary edition. Production was 2,050.

Window dressing aside, real history was made in March 1983 when Chevrolet finally introduced the all-new Corvette everyone had been waiting for— for nearly 15 years or roughly six months, depending on your perspective. The third-generation Corvette was a decade old when Dave McLellan began envisioning a redesigned next generation early in 1978. Initially it appeared his vision would become reality in 1983, but various stumbling blocks delayed that debut, which would have been

Even past 40,
America's sports car keeps rolling along in impressive fashion. ■

in the fall of 1982. While 43 pre-production third-generation 1983 Corvettes were built, none were released to the public, and only one still survives today. Chevrolet skipped over the 1983 model and introduced its next generation Corvette as a 1984 with an extended production run.

Everything about the car was new, from its modern chassis, to its roomier interior, and its re-styled body. Created by GM designer Jerry Palmer, the 1984 Corvette shell was state-of-the-art in both form and function. Its drag coefficient was .34, down nearly 25 percent in comparison to the body left behind in 1982. Beneath that shell was an innovative "bird cage" structure integrated with a "backbone-type" frame that mounted the drivetrain from engine to

LEFT
They called it the King of the Hill, and for good reason. Chevrolet's first ZR-1 was easily the most dominating street-going Corvette ever built. With a superb suspension, loads of rubber and 375 horses beneath its clamshell hood, the 1990 ZR-1 could run with anything in this country, and almost anything worldwide.

Like its forefather 35 years previously, all 1988 35th anniversary Corvettes were white. This external identification was included in the $4,795 package listed under RPO Z01.

Four cams, 32 valves and all-aluminum construction certainly qualified the ZR-1's LT5 V-8 as the most exotic production Corvette powerplant ever. Rated at 375hp from 1990 to 1992, the LT5 was pumped up to 405 horses for 1993-95.

The 40th anniversary Corvette's Ruby Red color scheme carried over into the interior. Headrests also included special "40th" logo embroidery.

differential as one rigid component joined by an aluminum C-section beam. Suspension was totally new, with fiberglass transverse monoleaf springs in the front and rear. Aluminum and other lightweight materials were used wherever possible to cut unwanted pounds.

Sixteen-inch cast-aluminum wheels measured a half inch wider in back. The four-wheel disc brakes had semi-metallic linings and aluminum calipers. The engine was a 205hp L83 Cross Fire 5.7-liter V-8. A choice was offered between a four-speed automatic or 4+3 Doug Nash manual (with overdrives in the top three gears). All this and more came standard with the 1984 Corvette. More than 51,000 were sold during the extended production run, kicking off the third generation in grand fashion.

Two years later, a convertible returned to the Corvette lineup after a nine-year hiatus. And it appeared just in time to pace the 70th running of the Indy 500 on May 25, 1986. For the second time, a collection of high-profile Corvette Indy Pace Car replicas was marketed to the public. A third Corvette would pace the lead lap at Indianapolis in 1995, spawning yet another group of Pace Car replicas.

Late in 1986, the first in a series of truly tough Corvettes appeared on the lot of a New Jersey Chevrolet dealership founded by Malcolm Konner. Konner Chevrolet put together 50 of its "Malcolm Konner Commemorative Edition" Corvettes, one of which was then retrofitted with a twin-tur-

Yet another birthday present to Corvette customers arrived in 1993, and exclusive paint was once more part of the package.

bocharged engine supplied by Callaway Cars, Inc., now of Old Lyme, Connecticut. Reeves Callaway, Callaway Cars founder, had been toying since 1977 with aftermarket turbo modifications of various models, from BMW to Volkswagen. But the Konner installation was just a stepping stone towards Callaway's biggest break—an agreement with Chevrolet to build the Callaway Twin Turbo Corvette.

Chevrolet officially assigned the Callaway package an RPO code, B2K, in June 1986. Production began the following month and the Twin Turbo killer Corvette debuted as a 1987 model. Chevrolet shipped fully assembled 1987 Corvettes to Callaway Cars where they were converted into Twin Turbos. Although the price for the "option" alone was $19,995, those in the need for speed couldn't have cared less. With 345hp, a 1987 Callaway Corvette could reportedly hit 177mph. One-hundred-eighty-four were built that first year. RPO B2K stayed on the Corvette options list through 1991.

With 345hp, a 1987 Callaway Corvette could reportedly hit 177mph. ■

Meanwhile, back in the regular-production world, Chevrolet again marked a fiberglass birthday in 1988 with a special 35th anniversary Corvette. Like the first Corvette in 1953, the 35th anniversary model came only in white. White leather sport buckets with anniversary headrest embroidery and a commemorative console plaque were also part of the deal. Power seats, air conditioning, a sport handling package, and external identification were included as well. Price for the package was $4,975; 2,050 were built.

The legendary "King of the Hill," the ZR-1, debuted as a 1990 model after the public was teased with a prototype introduction in 1989. The heart of this beast was the 375hp LT5 5.7-liter V-8, engineered by Lotus in

Callaway Cars' Twin Turbo Speedster of 1991 surely looked every bit as outrageously as it ran. Designed by Paul Deutschman, the Speedster's shell was best recognized for its cutdown windscreen and eye-popping paint schemes done in purple, yellow, green, orange or blue. Beneath those twin scoops was a 450hp 5.7-liter small-block fed by twin, intercooled turbochargers. Price for this beast was about $150,000.

Rarest of the 1993 40th anniversary Corvettes was the ZR-1 coupe. Only 245 were built. Even more special is this particular 40th anniversary ZR-1, which was modified by aftermarket builder Doug Rippie Motorsports. Only eight ZR-1s received DRM's emissions-legal "Black Widow" touch.

England and built by Mercury Marine in Stillwater, Oklahoma. Dual overhead cams, four valves per cylinder and an all-aluminum construction were just the beginning of the LT5's innovative appeal.

As for the rest of the package, ZR-1 brakes were huge—13 inches in front, 12 in the rear. And in order to house the equally huge Goodyear Eagle 315/35ZR-17 GS-C tires required in back to handle all that power, the ZR-1 received an exclusive widened tail section with unique square taillights. ABS and Z51 suspension were also included.

Whether in the curves or flat-out, the ZR-1 was a world-class street killer. Its time-honored 0-60 clocking was only 4.9 seconds. Quarter-mile performance came in at 13.4 seconds with a 108mph trap speed.

The DRM Black Widow LT5 got its name from its black finish. Thanks to, among other things, different cams and precise head work, the DRM LT5 produced 475hp. Also part of the Doug Rippie conversion was coilover shock front suspension.

Chevrolet introduced this eye-catching Copper Metallic finish for the Corvette in 1994, only to find it quite difficult to apply evenly. Because of this, the shade was discontinued early in the run. Reportedly, only 110 1994 Copper Metallic Corvettes were released, 86 coupes and 24 convertibles.

Superior sports car appeal carried over in typically stunning fashion for 1995. Improvements included better ride quality and the addition of larger (13-inch rotors in place of the 12-inch units used in 1994) heavy-duty standard brakes.

Top end was nearly 180mph. All this dominating power did, however, come at a price. ZR-1 equipment added around $25,000 to the $32,000 normal asking price for a 1990 Corvette coupe. While that figure didn't deter diehards at first, it did help limit the car's appeal. Actual production never did reach projections, not even after Chevrolet leapfrogged the 1993 ZR-1 over the Viper as the most powerful car in America by upping the LT5 ante to a whopping 405hp.

After 3,049 ZR-1 Corvettes were built the first year, production rapidly declined as the gleam wore off; even more so after Chevrolet introduced its second-generation LT1 in 1992. ZR-1 production dropped to 2,044 for 1991, 502 for 1992, and 448 for 1993, 1994 and 1995. According to Chevrolet general manager Jim Perkins, both the success of the much less expensive LT1 Corvette and the rising costs associated with continuing the ZR-1's limited production run helped spell the end for the King of the Hill.

After taking away a bit of the ZR-1's exclusive outward appeal by adding the

The legendary "King of the Hill," the ZR-1, debuted as a 1990 model after the public was teased with a prototype introduction in 1989. ■

widened tail to all Corvettes in 1991, Chevrolet really knocked the knees out from under the King of the Hill in 1992. New that year was a truly hot pushrod small-block, the second-generation LT1. Producing 300hp, the latest in Chevy's long line of 5.7-liter small-blocks went a long way toward tempting Corvette buyers to save their $25-30,000 and stick with standard Corvette performance. A 1992 LT1 could hit 60mph in 5.7 seconds, while the quarter-mile went by 8.4 ticks later. Top end was listed at a tad more than 160mph. LT1 muscle was just what the doctor ordered to revive the Corvette's spirits going into its fourth decade on the road.

To mark that special anniversary, Chevrolet once again rolled out a birthday edition. Wearing a price tag of $1,455, RPO Z25—Chevrolet's 40th anniversary appearance package—was available on all 1993 Corvettes, LT1

On August 28, 1995, the last of Chevrolet's 6,938 ZR-1 Corvettes rolled off the Bowling Green assembly line before an assembled crowd of company officials and press, signaling the end of the six-year run for the King of the Hill.

OSA
OSA
OSA
5.7L
V8 LT1
Y Car
3800

In its final form, the revered LT5 V-8 still pumped out 405hp, as it had since 1993. Adding those 30 additional horses was basically a matter of maximizing what already existed. Engineers simply ported and polished the cylinder heads.

coupes and convertibles, as well as the brawny ZR-1. Thrown in as part of the deal was exclusive Ruby Red metallic paint with matching leather inside, color-keyed wheel centers, 40th anniversary logos, and chromed emblems for the hood and fuel filler door. Additional special identification also appeared on the Ruby Red leather bucket seats in the form of "40th" logo headrest embroidery. Production for the Z25 1993 Corvette was 4,333 coupes, 2,171 convertibles and 245 ZR-1s.

Even past forty, America's sports car keeps rolling along in impressive fashion. Improvements to the LT1's induction gear helped the 1994 Corvette feel even stronger than the 1993, even though advertised output remained at 300hp. Minor overall improvements have followed, but how

LEFT
As in 1982, Chevrolet has put together a special Collector Edition model to honor an outgoing generation before the next new one arrives. The exclusive paint and interior treatment are part of the deal, as is the new 330hp LT4 V-8 if the buyer so chooses.

LEFT
The new LT4 small-block is standard for the 1996 Grand Sport, optional on all other Corvettes. Its 30 additional horses, compared to the standard 300hp LT1, will be warmly welcomed.

RIGHT
Painted to revive memories of its racing namesakes from earlier days, the 1996 Grand Sport is offered in coupe and convertible form. Plans call for about 1000 to be built, all identical externally.

much more can Chevrolet improve on what ranks as one of the world's best performance machines at the price? Just take a look.

New for 1996 is an optional 5.7-liter V-8, the LT4. Like its still-strong LT1 cousin, the LT4 has aluminum heads, roller lifters and sequential-port fuel injection. But various improvements, including a more aggressive cam, better breathing heads, upgraded, higher-flowing injection, and 10.8:1 compression (up from the LT1's 10.4:1), help bump output up from 300 horses to 330hp. Although it is available for all 1996 Corvettes, the LT4 is the standard powerplant for the Grand Sport special-edition models. Only the six-speed manual comes behind the LT4

Inspired by Duntov's five lightweight Grand Sport racers of 1963, the new Grand Sport features an exclusive competition-type paint scheme. All of the approximately 1000 Grand Sport coupes and convertibles planned for 1996 will be painted Admiral Blue Metallic with white racing stripes and red "Sebring-style" hash marks on the driver's side front fender. Various bits of special identification, 17-inch black aluminum wheels, black brake calipers with bright "Corvette" lettering, and special bucket seats are part of the Z16 package. Grand Sport coupes receive P275/40ZR-17 rubber up front and P315/35ZR-17 in the rear, with a pair of fender flares added in back to help house the wider tires. Grand Sport convertible tires are P255/45ZR-17 (front) and P285/40ZR-17 (back).

A second special-edition Corvette for 1996 also follows in another earlier model's tracks. Like its 1982 counterpart, the 1996 Collector Edition has arrived to salute the last of the latest generation of Corvettes before the next all-new model arrives. Exclusive Sebring Silver paint, special identification, silver 17-inch aluminum wheels, black brake calipers with bright "Corvette" lettering, and "Collector Edition" embroidery inside make up what Chevrolet calls an "eminently collectible" offering. As with base models, the LT1 is standard for the 1996 Collector Edition, with the LT4 available at extra cost.

Whether or not this newest Collector Edition Corvette will end up being eminently collectible is anyone's guess. Ask again a few more generations down the road.

429 RAM AIR
mach 1
MUSTANG
Radial T/A
BFGoodrich

FORD MUSTANG

Mike Mueller

Acknowledgments

Comparing the all-new '94 Mustang to its '64-1/2 ancestor on a beautiful fall day in Atlanta last year during one of the many Ford-sponsored celebrations of its latest ponycar rendition was only natural. As the ads proclaimed, "It was, it is." And with the photo appearing here, so began my quest to photograph as many Mustangs as possible for this short-winded epic. Hmmm, 30 years of Mustangs, where to begin?

Now that all that is done, the time comes to thank all the not-so-little people who made all my sweat possible. People like my brother, Dave Mueller, of Flat ville, Illinois, who never fails to come through for me when I'm on one of my many Midwest photo jaunts. Then there's sister Kathy Young, of Champaign, Illinois, my wife, Denise, and good friend Leslie Mathis, both here in Lakeland, Florida, all able-bodied Mustang pilots who each took turns behind ponycar wheels while I did all the snapping. And speaking of test drivers, I certainly can't forget little Teddy Mueller, who literally put the pedals to the metal in one of AMF's Midget Mustangs, owned by good sport Dale Richeson in Tuscola, Illinois.

Also from Tuscola, Mustang collectors and all-around great folks Tom and Ruthi O'Brien went well beyond the call of duty while allowing me access to their great cars. Of note as well is an equally great group of guys known as the Central Illinois Mustangers, Bernie Doty, Dennis Crow, Jim Fannin, Max Dilley and Dave Gass, to name just a few. Florida Mustangers Glenn Bornemann and Frank Cossota were of priceless assistance, too.

A big thanks also goes to Henry "Butch" Schroeder, of the Midwest Aviation Museum in Danville, Illinois, and his ace righthand man, Mike Vadeboncoeur, for letting me use Butch's fabulous F-6D P-51 Mustang fighter plane as a photo prop of all things. And I also can't thank noted Georgia musclecar collector Milton Robson and Wayne Allen enough, Wayne being Robson's highly talented restoration guru.

Other debts of gratitude are owed to Jim Sawyer and Dan Reid at Ford Special Vehicle Team in Detroit for the loan of not one but two '94 SVT Cobras. Tom Boyle, Anne Booker, and Barbara Kinnamon, Ford's public affairs people in Atlanta, also allowed the use of two '94 GTs. And an additional hot ride came courtesy of Dario Orlando of Steeda Autosports in Pompano Beach, Florida, who put me behind the wheel of his excellent '94 Steeda Mustang. Special thanks also go to Robert F. Tasca Sr. and his family, Robert Jr., Carl, David and Bobby III, in Rhode Island for their cooperation, patience and, above all, their hospitality.

Finally, this book could not have happened at all without the additional cooperation of the various feature car owners. Although I'd love to thank each one individually, space constraints permit only so much hot air in one blast, so one hearty thank you will have to do. In basic order of appearance, these people are:

Robert Deale III, Atlanta, Georgia, '64-1/2 289 convertible (red, this section); Max and Joyce Dilley, Urbana, Illinois, '67 GT coupe; Peter and Beverly Fagan, Urbana, Illinois, '67 GT fastback; Charlie and Pam Plylar, Kissimmee, Florida, '71 429 Cobra Jet Mach 1; Donald and Pam Farr, Mulberry, Florida, '66 GT coupe; Tom and Carol Podemski, South Bend, Indiana, '78 King Cobra; Mike and Denise Mueller, Mulberry, Florida, '88 GT convertible; Dale Richeson, Tuscola, Illinois, '64-1/2 260 coupe; Sam Munro, Dunedin, Florida, '64-1/2 289 convertible; Paul and Carolyn LiCalsi, Orlando, Florida, '65 289 2+2 fastback (red); David and Marilyn Gass, Rantoul, Illinois, '65 six-cylinder 2+2 fastback (blue); Bill and Janien Bush, Champaign, Illinois, '66 Hi-Po 289 convertible; Jim and Lynda Fannin, Bloomington, Illinois, '66 Hi-Po 289 GT coupe; Max and Joyce Dilley, Urbana, Illinois, '67 GTA convertible and '68 California Special; Tom and Ruthi O'Brien, Tuscola, Illinois, '68 fastback; Chris and Deborah Teeling, Enfield, Connecticut, '68-1/2 428 Cobra Jet fastback; Dennis and Kate Crow, Oakwood, Illinois, '69 428 Cobra Jet Mach 1; Milton Robson, Oakwood, Georgia, '69 428 Super Cobra Jet convertible; Bernie and Zona Gail Doty, Strasburg, Illinois, '70 Boss 302; Barry Larkins, Daytona Beach, Florida, '70 Boss 429; Steve and Stacy Collins, Jacksonville, Florida, '71 Boss 351; Ralph Gissal, Land O Lakes, Florida, '72 Olympic Sprint convertible; Tom and Carol Podemski, South Bend, Indiana, '73 convertible, '75 Mach 1 and '84-1/2 20th Anniversary hatchback; Jack McAllister, Dallas, Pennsylvania, '79 Indy Pace Car replica; Tom and Ruthi

O'Brien, Tuscola, Illinois, '83 GT, '83 5.0 convertible and '66 Shelby GT 350; J.D. Anderson, Vero Beach, Florida, '86 SVO; Dave Geiger, Elkhart, Indiana, '65 SOHC A/FX factory dragster; Dave Robb, Titusville, Florida, '67 Shelby GT 500; Dennis and Kate Crow, Oakwood, Illinois, '69 Shelby GT 350; Bill and Janien Bush, Champaign, Illinois, '69 Shelby GT 500 convertible; Bill and Barbara Jacobsen, Odessa, Florida, '70 Shelby GT 500; J. R. and Rita Castle, Clearwater, Florida, '90 ASC/McLaren; Ken and Mary Jean Wesche, Mulberry, Florida, '87 Saleen.

Again, thanks so much, all.

In the beginning Ford's Mustang represented affordable, practical sportiness, but that didn't mean ponycar buyers couldn't dress their mounts to the nines. Among the wide array of convenience, performance and appearance options initially offered in 1965 and '66 was this attractive styled-steel wheel. Shown here is the 1965 version.

1 Introduction

Thirty Years Down The Road And Still Kicking

They came from far and wide across this country, as well as overseas, automotive enthusiasts by the tens of thousands—including one president of the United States—all headed for Charlotte, North Carolina. The date was April 17, 1994. The reason for the mass exodus? A birthday party, a 30th anniversary celebration honoring the car that has probably turned more heads—among them Bill Clinton's—in its time than anything ever to roll out of Detroit. Trends, fads, and commander-in-chiefs have certainly come and gone over the last three decades, but Ford's Mustang has never left and remains running every bit as strong today as it did when it first hit the ground in a gallop in 1964. Thirty years on the job and still a winner—now that's something worth celebrating. Wonder if Mr. Bill took any notes?

Left: A ponycar progression. Shelby American's first GT 350 Mustang variant (back) is shown with Ford's original Mustang and two of its "prototype" predecessors, the mid-engined, two-seat Mustang I racer (front) from 1962 and the Mustang II showcar (directly above the Mustang I) of 1963.

Interestingly, Washington's Oval Office has been home to seven presidents since Lee Iacocca grabbed all the limelight immediately following the Mustang's official introduction on April 17, 1964. Kicking off a feeding frenzy the likes of which American automobile dealers had never seen before, Dearborn began rolling out those little affordable Mustangs with their long hoods, short rear decks, bucket seats, and floor shifters as fast as buyers could rope 'em in. Editors at *Time* and *Newsweek*

EAGLE ST
GOODYEAR

While many purists gasped, most among the automotive press were pleased with the Mustang's new body for 1967, a larger, more rounded shell created basically to make room for the breed's first big-block powerplant, the 390cid FE-series V8.

were so impressed with the "newest breed out of Detroit" they both immediately featured Iacocca and his hot-selling baby in prominent cover stories. Another high-profile honor quickly followed as the Mustang was chosen to pace the 48th running of the Indianapolis 500 in May 1964.

Twelve months later, Mustangs were still making news as production surpassed 417,000; that topped Detroit's existing record for first-year sales established in 1960 by Ford's Falcon. Yet another major milestone came in February 1966 when the millionth Mustang left the line. Hands down, no one has ever done it any better.

Of course Iacocca wasn't the only mover-and-shaker behind Dearborn's amazing mass-market marvel. Donald Frey, product planning manager; Hal Sperlich, his assistant; Donald Petersen, marketing man (later to become Ford's president); Gene Bordinat, styling chief; Joe Oros and Dave Ash, stylists; and many others promoted the roles in conception, design, promotion, and production of the Mustang. But it was up-and-coming Lee Iacocca as Ford vice president and division chief who stood ready, willing, and able to accept the lion's share of the credit.

Then again, only one man can be the father of anything, and if any one Ford man deserved that honor alone it had to be Iacocca, simply because the original idea of mass producing a small, affordable, youth-oriented automobile was his. And from there, it was his dedicated efforts to untie Henry Ford II's commonly knotted purse strings that singlehandedly transformed dreams into reality. As Donald Frey later recalled, "It took something like five shots with the senior officers of the company to convince them to put money into the car." Although various designers, engineers, and executives did nurture the idea, without Iacocca's influence atop Dearborn's ivory tower, the Mustang would've never been born.

Iacocca's idea began taking shape not long after he became division general manager in November 1960. By the summer of 1961, Gene Bordinat's Advanced Styling Studio had already put a small, sporty model on paper. In May 1962, the Mustang I project commenced. A topless two-seater with a midships-mounted four-cylinder engine, the Mustang I was certainly intriguing, but it didn't mesh with Iacocca's plans for low-priced, wide-appeal, exciting-yet-practical transportation. "Up until that point," remembered Frey, "we had been thinking two-seaters. But [Iacocca] was right; there was a much bigger market for a four-seater."

Iacocca's challenge to his styling staff that summer to create a youthful four-seater produced rapid results. Although studio head Joe Oros is commonly credited with the shape, it was his assistant, Dave Ash, who designed the so-called "Cougar," a long-hood/short-deck styling proposition that went from clay to regular production in short order, with relatively few modifications. Familiar mockups were appearing by August 1962, followed in October 1963 by the prophetic Mustang II, a four-place show-car that demonstrated in no uncertain

If curbside critics thought the 1967 ponycar redesign was big news they had another thing coming in 1971 when the Mustang grew, somewhat alarmingly, on all fronts.

Compare this Grabber Yellow '71 Cobra Jet Mach 1 to its Signalflare Red '66 GT forerunner.

Long disappointed with the Mustang's direction after 1966, Lee Iacocca attempted to turn things around in 1974 by rolling out the little Mustang II. But while initial responses bordered on ecstatic, the idea quickly soured. And by 1978 the second-generation ponycar was history. Most prominent of this short-lived bloodline was the spoilered and striped '78 King Cobra.

terms that Ford had given up on the two-place ideal originally demonstrated by the Mustang I. Many among the sporting public were not pleased, however, especially those at *Motor Trend* who felt it was "a shame the Mustang name had to be diluted this way."

Such thoughts were quickly swept away by the flood of positive responses to Ford's all-new Mustang, introduced at the New York World's Fair as a midyear model, later to become known as a "1964-1/2" offering. *Car and Driver* called it "the best thing to come out of Dearborn since

Mustang performance really took off in the mid-'80s after more than a dozen years of Camaro dominance on Mainstreet U.S.A. A minor GT restyle in 1987 supplied the "aero" look while 225 horses worth of 5.0L V8 backed up that look with some serious muscle. This red GT convertible is a 1988 model. Other than a few minor changes—primarily wheels and tires—the Mustang exterior remained essentially unchanged up through 1993.

After rolling on the same "Fox chassis" platform for 14 years, Ford redesigned and totally restyled the Mustang for 1994. Underneath that attractive, modernized shell is, among other things, an exceptionally rigid chassis and standard four-wheel disc brakes. GT power still comes from the ever-present 5.0L small-block V8, now rated at 215 horses.

"The best thing to come out of Dearborn since the 1932 V-8 Model B roadster."

-Car and Driver

the 1932 V-8 Model B roadster." In the words of *Road & Track's* Gene Booth, the Mustang was "a car for the enthusiast who may be a family man, but likes his transportation to be more sporting."

Imagine that, a youthful, sporty flair in a practical, easy-to-handle package, all for around $2400. Bucket seats and floor shifter standard. Crisp, fresh styling that belied the cost-saving Falcon-based platform beneath the skin. Loads of options. Even more performance potential. Any way you looked at it, the new Mustang was many things to many people, some 417,000 people to be semi-exact. In base form with its economical six-cylinder engine, it was a practical, budget-minded grocery-getter that brought home the bacon in style. Go-getter was a better term for a Mustang armed with one of the hot optional V8s, which would only get hotter each year.

While power sources grew stronger, the Mustang grew as well, basically to make more room for more engine. A restyle for 1967 produced a wider, rounded body, but that was nothing compared to the longer, heavier, palatial pony introduced for 1971. A large look that carried over into 1973 as sagging sales signalled the end for the first-generation Mustang. What followed was Iacocca's attempt to restore his original ideal, and the downsized Mustang II initially sold exceptionally well, then it too faded, lasting just five years through 1978. A totally redesigned "Fox-chassis" Mustang appeared the following year, kicking off the third generation, which managed to stay on the scene until 1993. And now we have yet another generation of Mustang, the totally redesigned SN-95 platform. What it was, it is, you can bet your spurs.

Thirty years and still running, how much longer is anyone's guess. One thing's for certain, though. Rivals and imposters may come and go, but there will always be only one Mustang, the progenitor of the long-hood/short-deck breed. Camaro, Firebird, you name it; regardless of the marque, there still all known as "ponycars."

As it did in 1964 and 1979, the latest, greatest all-new Mustang was chosen as the official pace car for the Indianapolis 500. Performing that duty at The Brickyard in May 1994 was a Special Vehicle Team Cobra, a car powered by a modified 240hp version of the 5.0L V8. This '94 SVT Cobra droptop was one of 1000 Indy Pace Car replicas built to mark the occasion. Adding the typical official decal to the door was up to the owner.

Illinois
MY HIPO

2 1964-1966

Off And Running On The Inside Track

While Ford's Mustang is credited with kicking off the ponycar breed, it wasn't exactly the first entrant in Detroit's long-hood/short-deck derby. That honor actually belongs to Plymouth's Barracuda, an equally small, sporty machine that beat the Mustang out of the gates by a couple weeks in April 1964. Basically a Valiant with a large, sloping rear window grafted on; the hastily created Barracuda didn't quite turn heads the way the Mustang did with its totally fresh image, a fact quickly demonstrated by the runaway lead Ford took in the early ponycar sales race. As total Mustang production was zooming past 500,000 into the car's second year, Plymouth was moving 23,443 '64 Barracudas, followed by another 64,596 in 1965.

Of course more competition would come soon enough. By late 1966, General Motors was rolling out its own ponycars, Chevrolet's Camaro and Pontiac's Firebird, while Mercury was letting its luxurious Cougar run free in the Ford Motor Company corral. A year later, even American Motors was entering the race, introducing its Javelin and two-seat AMX as 1968 models. But for nearly the first two and a half years of its life, Ford's wildly popular Mustang basically ran unopposed, a situation Iacocca had planned all along.

Ford's all-new breed debuted on April 17, 1964, as a "notchback" coupe and a sexy

Topless travel has always represented the only way to fly for the sporty set, and Ford Mustang buyers were no exception in 1966. Throw in a set of optional styled-steel wheels and the hot 271hp "Hi-Po" 289 V8 and the deal was irrestible. Notice this rare Hi-Po convertible is not a GT.

Ford's Mustang was by no means the first to demonstrate the long-hood/short-deck theme—consider Chevrolet's Corvette, Studebaker's Avanti and Ford's own two-seat Thunderbirds of 1955-57—but may well stand today as the car best recognized for fitting the profile to a T. While all first-year Mustangs sold between April 1964 and August 1965 were originally referred to as 1965 models, the early run ending in August 1964 was soon labeled "'64-1/2." This '64-1/2 coupe features Pagoda Green paint, one of five shades only offered on those early Mustangs. Also notice the "260" fender badge. The optional 164hp 260cid 2V (two-barrel carburetor) V8 was traded for a 200hp 289cid 2V Windsor small-block after the so-called '64-1/2 production ended.

convertible. An idea basically "borrowed" from Chevrolet's Corvair Monza, the standard combination of bucket seats and a floor shifter established the Mustang's sporty feel, regardless of which engine was beneath that long hood. And although a Falcon-style instrument panel served as a reminder that much of the structure beneath the Mustang's skin was passed up from Ford's little econo-buggy, most who took a seat behind that racy three-spoke steering wheel couldn't have cared less.

Desirable options, both factory-supplied or dealer installed, were plentiful,

Helping enhance the Mustang's sporty image were standard bucket seats and a floor shifter. A console was optional, as was underdash air conditioning. Integral in-dash air conditioning wouldn't be incorporated into the design until 1967. Base Mustangs (non-GTs and those without the optional deluxe Interior Decor group) built prior to 1966 can also be identified by their simple Falcon-based instrument panel.

including typical amenities like power steering and brakes, underdash air conditioning, a power top for convertibles, and Ford's three-speed Cruise-O-Matic automatic transmission. Sport-minded buyers could've also added a four-speed stick, an interior console, the popular "Rally Pac" (a column-mounted tachometer/clock combo), a special handling package (heavy-duty suspension), and the attractive styled-steel 14-inch five-spoke wheels. Also helping dress things up outside was a vinyl roof and various simulated wire wheelcovers.

Base power came from a 101hp 170cid six-cylinder powerplant, replaced that fall by a stronger, more durable 200cid six, rated at 120 horses. Optional V8s included the Falcon's 164hp 260 with a two-barrel carburetor and its slightly larger Windsor small-block brother, the 210-horse 289, fed by a four-barrel. The 260 V8 was soon dropped, replaced by a 200hp 289 two-barrel once the so-called '64-1/2 Mustang gave way to "true" 1965 production in August 1964. A higher-compression 225hp 289 four-barrel V8 was also added, superseding the original 210hp 289.

Listed in April 1964 but not available until June 1, the truly hot High Performance 289 joined the Mustang lineup late, then instantly hit the ground running. Special heads, a potent solid-lifter cam, a 600-cfm Autolite four-barrel breathing through a chrome open-element air cleaner, beefed main bearing caps on the block's lower end, a mechanical-advance distributor, and free-breathing exhaust manifolds helped set the "Hi-Po" 289 apart from the herd. Compression was 10.5:1 while the all-important output figure was advertised at 271 horsepower.

The Hi-Po 289 was just the kick the new ponycar needed. *Road & Track's* crit-

28 - 2300
FLORIDA

Left: All '64-1/2 Mustangs were either coupes or convertibles as the 2+2 fastback didn't join the ponycar ranks until after "actual" 1965 production began. This '64-1/2 droptop is powered by the optional 210hp 289 4V V8, initially the hottest Mustang powerplant available until the 271hp High-Performance 289 was unleashed in June 1964.

ics called the 271hp Mustang a "four-passenger Cobra." Equally impressed were *Car Life's* testers, who lauded the Hi-Po for its "obvious superiority to the more mundane everyday Mustang." Continued *Car Life's* September 1964 report, "where the latter has a style and a flair of design that promises a road-hugging sort

When introduced in April 1964, the Mustang offered three power choices: the standard 101hp 170cid six-cylinder and two Windsor small-block V8s, the 164hp 260 2V and this 289 4V, rated at 210 horsepower. The optional 271hp "Hi-Po" 289 appeared in June, while the three original engines were replaced later that fall; the 170 six by a 200cid version, the 260 V8 by a 200hp 289 2V, and the 210hp 289 by the 225hp 289 4V. Notice the generator (just above the battery) on this 210hp 289; it represents one of the easiest ways to identify a '64-1/2 Mustang. Later models used alternators.

Pedal Cars

Feet Don't Fail Me Now

Frenzied car buyers weren't the only ones to notice Dearborn's enormously popular Mustang in the spring of 1964. About the time Iacocca and crew were rushing their ponycar project to market, toymakers at the American Machine and Foundry Company (AMF), in Olney, Illinois, were putting together their own plan to jump on the Mustang bandwagon—with both feet. Presented to Iacocca by AMF's Patrick Wilkins were three prototype vehicles, truly small cars intended for truly youthful drivers. Ford's vice president liked the idea, and thus the Mustang pedal car was born.

Sold through Ford dealers, AMF's "Midget Mustang" pedal car appeared just in time for Christmas 1964. Full-page ads in at least a half dozen major magazines announced the Midget Mustang's debut to the kids while also enticing mom and dad to put themselves behind the wheel of a somewhat larger, fossile-fuel-fired ponycar convertible. Price for the pedal-driven variety was $12.95.

Mustang pedal cars stood only 14 inches tall, rolled on a 23-inch wheelbase, and measured 39 inches front to rear. Whitewall rubber tires were standard, as were deluxe wheelcovers and a steel three-spoke steering wheel, items all modelled after the real things. Additional sporty features included a Rally-Pac instrument cluster decal and a workable three-speed stick mounted atop the righthand door. Although rumors claim a few Midget Mustangs may have been painted light blue/grey, all Ford-marketed pedal cars were red, at least originally. Many were soon repainted to match dad's full-sized Mustang.

Introduced just in time for Christmas 1964, AMF's Midget Mustang pedal cars were sold through Ford dealers up through 1965. Although rumors of light blue examples exist, all were apparently red with rubber tires and a stickshift on the passenger door. Many were later repainted to match dad and mom's full-sized ponycar.

Another variation involved the limited-edition Indianapolis 500 Pace Car pedal cars distributed earlier to the Indy 500 Festival Committee. Appropriately painted white with a blue stripe down the hood and decklid and wearing bodyside decals, these pedal-powered pacers featured a hand-lettered serial number tag beneath the skin. Reportedly, only 100 were produced, with maybe four still known today, one bringing a then-record $1,700 at a recent collector auction.

Ford dealers continued selling the Midget Mustangs through the 1965 Christmas season, then Dearborn decided to drop the promotion in 1966. AMF wasn't deterred, however, and continued marketing the Mustang pedal car—sans official Ford Motor Company logos—on its own before ending production in 1972. Various minor changes were made during that span, the most notable being a switch from red paint to yellow in 1971.

According to toy expert Ken Schmidt, owner of Indianapolis' Blue Diamond Classics, a leading parts source and restorer of Mustang pedal cars, AMF built 110,812 kid-powered ponycar convertibles, all using the same stamped steel body modeled after the original Mustang. After ceasing production in 1972, AMF sold its tooling to the CIA Corporation in Mexico City, Mexico, where Mustang pedal cars production resumed, this time in various colors. Another attempt to revive the breed came in 1984 when The Little Car Company of San Diego attempted to import 1000 pedal cars from the Mexican firm to help commemorate the Mustang's 25th anniversary. Poor quality, however, hindered the effort and only 200 were actually distributed among U.S. dealers.

Whether built in Mexico or Illinois, all Mustang pedal cars are valued collectibles today. Schmidt claims a CIA-marketed car, new in its box, is worth about $650, while pristine AMF-produced examples have sold for twice that price. Blue Diamond Classics asks as much as $1000 to restore well-used Midget Mustangs. Clearly, playing with Mustang pedal cars can no longer be considered a kid's game.

That $12.95 asking price wasn't exactly cheap for a toy in 1964, but it pales in comparison to the amounts paid by collectors today, totals sometimes surpassing $1000. AMF continued manufacturing its Mustang pedal cars even after Ford dropped support after 1965. Later models were devoid of official Ford identification and were painted yellow beginning in 1971. Production ended the following year.

Easily the most valuable of the Mustang pedal car herd is this Indy Pace Car replica, one of maybe 100 specially serial-numbered toys distributed among the Indianapolis 500 Festival Committee in May 1964. Only about four of these are known today, according to Blue Diamond Classics' Ken Schmidt of Indianapolis, perhaps this country's leading Mustang pedal car restoration expert.

of performance and then falls slightly short of this self-established goal, the HP Mustang backs up its looks in spades." Quoted quarter-mile time for a Hi-Po Mustang with a 3.89:1 differential was 15.9 seconds at 85 mph. *Car and Driver's* hotfoots pushed the envelope even further with a 271hp Mustang motivated by 4.11:1 rear gears, managing a sizzling 14-second quarter-mile pass at 100 mph.

Three months after the impressive Hi-Po 289 appeared, Ford enhanced the

Also introduced in the fall of 1964 was a third Mustang bodystyle, the sexy 2+2 fastback. This '65 2+2 features upscale optional wire wheelcovers. But since its fenders wear no badges, underhood power comes from a budget-minded base six-cylinder.

Mustang's sporty image another notch with a third bodystyle, the "2+2" fastback, officially introduced September 9, 1964. With its seductive, sweeping roofline and similarly effective fold-down rear seat, the Mustang 2+2 was an instant crowd favorite, drawing more than 77,000 buyers in 1965.

Ponycar customers who wanted even more pizzazz, performance, and prestige were surely pleased with two more optional packages introduced in April 1965 to mark the Mustang's one-year anniversary. The interior decor group, more affectionately now known as the "pony interior" for the running horses embossed into the seat inserts, spruced up things inside considerably. Along with those intriguing seat inserts—used both front and rear—the pony interior also included luxury door panels, a simulated walnut rim for the steering wheel, pistol-grip door handles, a five-dial instrument panel, various bright mouldings, more simulated walnut here and there, and door-mounted red-and-white courtesy lights. Although the interior decor option was available for all three bodystyles, a fastback received a slightly different version as the fold-down rear seat did not have the pony inserts.

While the pony interior wowed 'em inside, the other April 1965 addition to the Mustang options list turned heads in more ways than one. Featuring both a hefty dose of eye-catching exterior imagery and a heaping helping of heavy-duty hardware, the GT equipment group was just the ticket for those among the ponycar faithful who wanted it all—sporty looks and gutsy performance. Available with the 225- and 271-horse 289 V8s only, the GT group included power front disc brakes and the special handling package, which added stiffer springs, bigger shocks, quicker 22:1 steering (standard steering was 27:1), and a thicker front stabilizer bar. GT Mustangs were

This larger, beefier 120hp 200cid six-cylinder replaced the original 101hp 170 six in the fall of 1964. Revised features included seven main bearings, a new cam, bigger valves, and higher compression, 9.2:1 compared to 8.7:1. Base six-cylinder sales made up 35.6 percent of 1965 Mustang production.

Previous page: Along with special "2+2" fender emblems, '65 Mustang fastbacks also received chrome rocker mouldings as standard equipment—they were optional for coupes and convertibles. Also notice the lack of trim within the simulated rear-quarter vent. Fastback and GTs in 1965 and '66 didn't have this trim, while coupes and convertibles did. Optional styled-steel wheels for 1965 were fully chromed while the same type sport wheel in 1966 featured only a bright center section with a chromed trim ring covering the unplated rim.

Left: Beneath that sexy, sweeping roofline, a 2+2 Mustang featured a practical fold-down rear seat and trunk space access from the interior, an arrangement that opened up all kinds of possibilities whether hauling fence posts or attending submarine races.

also adorned with appropriate fender badges, lower bodyside stripes, grille-mounted fog lamps, and twin exhaust trumpets exiting through the rear valance. The optional four-speed, limited-slip differential, and styled-steel wheels would've been icing on the cake as far as supreme GT performance was concerned in 1965 and '66. Was there a better way to fly?

Perhaps not. Then again, not all Mustangers looked to the sky. Budget-conscious, economical ponycars were far more plentiful than their high-flying, high-profile GT cousins, and this plain truth wasn't overlooked by Dearborn's gay marketeers.

As part of an early 1966 sales push centered around the upcoming production of the millionth Mustang, Ford put together yet another special ponycar, this one geared towards promoting the yeoman six-cylinder models. Offered as a hardtop convertible and rarely seen fastback, the Sprint 200 Mustang featured bodyside accent pinstripes, wire wheelcovers, an interior console, and a chrome air cleaner with an appropriate decal atop the 200cid six beneath the hood. Sure, it wasn't all that awe-inspiring, but the Sprint 200 did demonstrate that practicality didn't necessarily have to be dull, something Mustang dealers already knew.

After rolling out nearly 700,000 Mustangs during the extended 1965 model run beginning in April 1964, Ford followed up with some 607,000 more in 1966. Not wanting to break something that didn't need fixing, Dearborn designers barely touched their second-edition ponycar, letting the same shell return in 1966, showing only minor trim changes here and there. With demand being so great, who cared?

Left-bottom: Changes for 1966 were minor, to say the least, as all Mustang sheetmetal and bumpers interchanged with their 1965 counterparts. A new grille with chrome horizontal strips and a fully "floating" running horse logo, a restyled gas cap, and the addition of three horizontal strips to those fake rear-quarter vents were basically it on the outside. Also, compare the optional 14-inch styled-steel wheels on this "Hi-Po" '66 convertible to the '65 2+2's wheels in this chapter.

Optional equipment on this '66 coupe includes the black vinyl top, styled-steel wheels, GT equipment group and High-Performance 289 V8. Introduced in April 1965 to help mark the Mustang's first anniversary, the GT package included a nice array of image and performance pieces, including twin foglamps in a blacked-out grille, "GT" fender emblems, lower bodyside stripes, front discs, heavy-duty suspension and dual exhausts exiting through cutouts in the rear valance. Adding the optional K-code "Hi-Po" 289 only helped sweeten the GT pot even further. Hi-Po Mustangs received standard redline tires unless superseded by optional whitewalls, as is the case with the red 271hp convertible shown in this chapter.

Right: Rated at 271 horsepower, the Hi-Po 289 was the hottest Mustang powerplant available from its debut in June 1964 up until the 390cid big-block V8 appeared as a Mustang option in 1967. Although still offered that year, few 271hp 289s found their way beneath '67 Mustang hoods. As in 1965, the '66 Hi-Po shown here features chrome dress-up and an open-element air cleaner. Special valvetrain gear and beefier connecting rods and main bearing caps made the Hi-Po 289 a high-winding demon. Maximum power came on at 6000 rpm and 7000 revs were no problem.

Introduced along with the GT equipment group in April 1965, the Interior Decor package added deluxe appointments inside, including these attractive "running horse" seat inserts. Called the "pony interior," this option carried over almost unchanged into 1966, with the main difference being the five-dial instrument panel included as part of the deal in 1965. It was already standard on all models the following year.

Margaret
mach 1
GOODYEAR
SCJ 1969

3 1967-1970

From Polite Pony to Bucking Bronco

Perhaps the best combination of ponycar car pizzazz and performance, the Mach 1 Mustang debuted to rave reviews in 1969. "Are you ready for the first great Mustang?" asked a Car Life *report. "One with performance to match its looks, handling to send imported-car fans home mumbling to themselves, and an interior as elegant and livable as a gentleman's club?" Inside the Mach 1* "SportsRoof" *were high-back buckets, a sport steering wheel, a console, and lots of simulated woodgrain. Outside were various dress-up features, like a blacked-out hood, non-functional scoop, racing-type hood pins, color-keyed racing mirrors, a pop-open gas cap, and chromed styled-steel wheels. Engine choice varied; this '69 Mach 1 is powered by the top-dog 335hp 428 Cobra Jet with functional ram-air, a powerful package representing suitable symmetry for the equally intimidating North-American P-51 fighter plane in the background.*

Like many ponycar purists then and now, Lee Iacocca never did approve of the new direction taken by his pet project in 1967. "Within a few years after its introduction," he later wrote in his autobiography, "the Mustang was no longer a sleek horse, it was more like a fat pig."

Indeed, Iacocca's once-petite baby did grow some in time for the 1967 model year. Length stretched 2in, width increased by more than 2.5 inches, and height went up a 0.5in. While the new model's wheelbase remained at 108 inches, both front and rear tracks were widened to 58 inches, from 55.4 and 56, respectively. And of course all this up-sizing meant more weight too, about 130 lbs worth. Exterior styling swelled as well; in place of the earlier crisp, light lines was a bulging, rounded facade.

Lee Iacocca, the so-called father of the Mustang, didn't like the 1967 restyle at all, nor did he approve of the prime motivation behind rounding and widening the ponycar's flanks—to make room for the big-block 390 GT V8. Nonetheless, the restyled look was certainly not lacking in attractiveness. Even with the new concave taillight cove, twin-vented hood and revised measurements, much of the original Mustang image carried over. And again, GT models represented the top of the ponycar heap.

Obesity, however, is in the eye of the beholder. Not everyone agreed with Iacocca, and no one in 1967 was by any means ready to write the Mustang off as a great idea gone bad. Detractors, in fact, were in the minority as buyers continued flocking to Ford dealers in droves, inspired in part by the soothing words of *Hot Rod's* Eric Dahlquist. "Detroit has cobbled up so many fine designs in the last 20 years that when Ford decided to change the Mustang, everybody held their breath," wrote Dahlquist in *HRM's* March 1967 issue. "But it's okay, people, everythin's gonna' be all right."

As for why Ford had decided to change the Mustang, to fix something that on the surface didn't appear to need fixing, the answer involved rolling stones. Among others, Hal Sperlich, Donald Frey's special projects assistant, didn't want to see Ford's all-new ponycar gather moss. Even as the first Mustangs were being bought at a record pace in 1964, Sperlich was looking ahead to what everyone hoped would be a bigger, better model.

As chief Mustang product planner Ross Humphries later told author Gary Witzenburg in 1978, "Hal's feeling was that in the past when we had brought out a new car that was a winner, we had rested on our laurels and didn't do enough to upgrade the car to keep the momentum going. And soon our friends across town would come back and do us one better. Hal's philosophy on the '67 Mustang was to one-up the original in every respect without making a major change."

In order to fit the big 390cid FE-series big-block V8 beneath a '67 Mustang's long hood, designers had to rearrange the shock towers to supply enough side-to-side clearance. Fed by a four-barrel carburetor, the 390 GT V8 was rated at 320 horsepower. GT Mustangs in 1967 could've been equipped with the 390 big-block or 289 small-block.

Engineering changes beneath the skin involved improving ride and handling by revamping the basic platform and front suspension, which was also widened to make room for more engine beneath the hood. Basically, it was the "more engine" that made for more Mustang in 1967.

A new option for 1967 was this tilt steering wheel that, along with adjusting to nine different driving positions, would automatically swing up and away towards the center of the interior once the driver's door was opened. Also notice the optional in-dash air conditioning (vents above the radio) and stylish full console with storage compartment, both new for 1967. Optional speed control was introduced that year as well.

Mechanical parameters and sheetmetal simply had to expand to allow engineers to drop Ford's FE-series big-block V8 into the Mustang, a move that more or less had to happen if Dearborn wanted to stay at the head of the ponycar pack.

Although somewhat slow in coming, General Motors' ponycar spin-offs finally appeared in late 1966, and both Chevrolet's '67 Camaro and Pontiac's '67 Firebird hit the ground running hard, each with an optional high-powered big-block V8, Chevy's 396 and PMD's 400. To keep up, Ford let loose its first big-block Mustang, powered by the 320-horse 390 FE V8, offered with or without the GT equipment group. Also listed for '67 Mustang GTs only was Ford's former top-dog ponycar powerplant, the 271hp Hi-Po 289 small-block, making its final appearance.

A new twist for 1967 involved Mustang GT transmission choices. Choosing the optional Select-Shift Cruise-O-Matic automatic transmission added a red-accented "A" to the GT emblem. All "GT" models were equipped with manual transmissions, while all "GTA" Mustangs had automatics. Although a neat little identification trick, the GTA did not carry over into 1968.

In base form, the new-look '67 Mustang remained a budget-conscious sportabout with six-cylinder power. But Ford promotional men were more than proud of their equally new power lineup, now made up of 13 different engine/transmission combi-

An identification trick used only in 1967 involved the GT Mustang's transmission. While a manual-trans '67 GT was simply a GT, those equipped with the Select-Shift Cruise-O-Matic automatic received a red "A" on their fenders. All GTs in 1967 had sticks, all automatic cars were "GTA" models. This GTA convertible is powered by a 225hp 289 4V V8.

nations. Six-cylinder, small-block or big-block, three-speed, four-speed or automatic; it was in there. So were various structural enhancements and new options like a tilt-away steering wheel, cruise control, a folding glass convertible top window, and fully integrated in-dash air conditioning. A one-up on the original? *Motor Trend's* John Ethridge thought so, claiming "just about every Mustang change for '67 seems to bring the car closer to our notion of a true Grand Touring machine."

From a top performance perspective, however, not all was well in 1967. Even with the 390 big-block, a '67 Mustang GT was still no match for a 396 Camaro SS or Firebird 400. Using a fast-food analogy in reference to the 390 GT, *Car Life's* curbside critics suggested that "perhaps this super-burger, if it is to be a superburger, needs a little more mustard."

Ready with a jar of French's was renowned East Coast Ford dealer Robert F. Tasca of East Providence, Rhode Island.

Among noticeable exterior revisions for 1968 was different grille trim, "Mustang" fender script instead of block fenders, revised trim again for those fake rear-quarter vents, and totally redesigned optional styled steel wheels. When included along with the GT equipment, those chromed wheels wore center caps with large "GT" identification.

One of the country's most prominent promoters of Blue Oval performance, Bob Tasca was certainly no stranger to making Fords go fast, both at the dragstrip and on the street. But while Ford's so-called "Total Performance" campaign, originally initiated in 1963, had produced a worldwide race-winning reputation for Dearborn, regular-production performance available to Tasca and other dealers quickly fell well behind the competition. Tasca began noticing the shortfall in 1966.

"We did well from '63 to '65, when the car-buying market was a young one," Bob Tasca told *Super Stock* in 1968. Then "the younger people [became] disenchanted with Ford's performance on the street, and stopped buying." Reportedly, Detroit sold 634,434 high-performance cars (those with more than 300 horsepower) in 1966. Ford's

Ford put the Mustang on the musclecar map in April 1968, introducing the fabled 428 Cobra Jet variety. With 335 CJ horses between its flanks and the complete array of GT equipment throughout, the '68-1/2 Cobra Jet Mustang was a winner on the street, while on the run or standing still. This CJ Mustang was originally sold at Tasca Ford in East Providence, Rhode Island, the "birthplace" of the 428 Cobra Jet. It was Bob Tasca's men who in 1967 first began experimenting with mixing and matching Ford FE-series big-block parts to make the Mustang more competitive on the street performance scene.

piece of that performance pie? A mere 7.5 percent. "Shameful for a 'Total Performance' company," said Tasca.

Ford's first big-block Mustang did little to ease that shame. As Tasca Ford performance manager Dean Gregson told *Hot Rod's* Dahlquist in late 1967, "we sold a lot of 390 Mustangs last fall and into the winter, but by March they dropped off to practically nothing. That's when the snow melted off the asphalt." Continued Gregson, "we found the car so non-competitive we began to feel we were cheating the customer."

The 335hp Cobra Jet was based on a 428 passenger car block but included 427 low-riser heads, a cast-iron version of Ford's aluminum Police Interceptor (PI) intake mounting a big 735cfm Holley four-barrel, a 390 GT cam, PI rods, 10.6:1 pistons, low-restriction exhausts, and functional ram-air. Power front discs, braced shock towers, staggered rear shocks (on four-speed models), and a beefy 9-inch rearend were also included in the deal.

That summer, Tasca had taken it on himself to make the new big-block Mustang more competitive using parts right off the Ford shelf. Instead of the 390 FE, Tasca's men tried a 428 Police Interceptor V8 using big-valve heads and a 735-cfm Holley four-barrel. Presto, an instant 13.39-second quarter-mile sizzler. Soon after seeing Tasca's so-called "KR" Mustang—KR for King of the Road—Ford officials decided to build a regular-production counterpart, exchanged the KR designation for "Cobra Jet," and introduced the 428 CJ Mustang on April 1, 1968. Conservatively rated at 335 horsepower, the Tasca-inspired 428 Cobra Jet appeared just in time to save Ford's bacon on the street performance scene.

Looking much like one of Carroll Shelby's variants, the '68 California Special Mustang was a special promotional model put together by the Southern California Ford Dealers using many Shelby components. Offered only in GT coupe form with a six-cylinder as standard, the GT/CS featured a blacked-out grille with Lucas foglamps, twin hood locks, non-functional Shelby rear-quarter scoops, various striping, and "California Special" script in back. All engines were optional for the California Special.

TASCA

Win On Sunday, Sell On Monday

Any Ford performance fan worth his salt knows the name Tasca. Tasca Ford, owned and operated by Robert F. Tasca Sr., was a mecca for East Coast Blue Oval horsepower hounds in the '60s. If it was hot and powered by Ford, you could've found it at 777 Taunton Avenue in East Providence, Rhode Island, from Cobras to Shelbys to 427 Galaxies to Mustang GTs. And parts, too. If it made Fords go fast, Bob Tasca—"the Bopper"—sold it.

Tasca Ford also raced it. The Bopper may not have been the first to recognize the relationship between sizzling victories at the track and hot sales on the street, but he was the man who coined the phrase, "win on Sunday, sell on Monday." Tasca Ford did both in spades, beginning in 1962 when the dealership's special competition department was formed, headed by Dean Gregson. Ace mechanics included John Healy and Ralph Poirier, who still works for The Bopper today.

Driver Bill Lawton joined the team soon afterward and drag racing victories quickly followed. Pilot-

Robert F. Tasca today at his home in Rhode Island with his 1930 Model A, a present from his three sons on his 65th birthday in October 1991. Once the "mascot" car at Georgia Tech, Tasca's Model A was restored at Ford's Atlanta plant.

The Tasca tradition continues—Bobby Tasca III stands proudly with his '92 Mustang, also a birthday present, in this case for the younger Tasca's 16th anniversary. Specially prepared both on the Ford line and at Jack Roush's shop, Tasca's GT is powered by a 400hp 340cid small-block put together at Tasca Lincoln-Mercury in Seekonk, Massachusetts. At the track, Bobby Tasca's Mustang will do 12.2 seconds in the quarter, topping out at 113mph.

ing Tasca Ford's '64 Thunderbolt, Lawton won top Super Stock honors at the National Hot Rod Association's 1964 Winternationals in Pomona, California. Lawton won again at the 1965 NHRA Winternationals, this time in Tasca's '65 A/FX Mustang. Between 1962 and '68, Bob Tasca spent nearly $300,000 racing, a wise investment considering the national exposure helped his dealership grow into this country's second leading Ford dealer.

Although a strong racing presence did greatly contribute to that rise, so too did Bob Tasca's marketing expertise, his attention to details and his drive to be the best. A "grease monkey" at age 17, Tasca had first gone to work in May 1943 for Sandager Ford in Cranston, Rhode Island. By 1948, he had climbed the ladder to sales manager, then to general manager the following year. In November 1953, he opened his own dealership, making customer satisfaction his top priority, as it still is four decades later. As one Tasca employee put it, "Bob Tasca will do whatever it takes, whatever the cost to satisfy a customer—he will not quit until that guy leaves here happy."

Today, Bob Tasca's office is in Seekonk, Massachusetts, just a few blocks east of where the old Tasca Ford dealership used to be before it closed down in 1971. No, Tasca didn't go out of business, he just changed gears, trading Fords for Lincolns and Mercurys. And of course, not one Lincoln-Mercury dealer in the world does it better than Seekonk's Tasca Lincoln-Mercury, run by a second generation of Tascas, sons Robert Jr, Carl and David.

It's somewhat ironic. Robert F. Tasca Sr. first got his Rhode Island dealership rolling around 1960 by dressing up Fords with special custom luxury features. Once horsepower became the going thing, he began using that tool to promote high-powered sales, during which time he was directly responsible for Ford Motor Company's introduction of its street performance savior, the Cobra Jet Mustang, in April 1968. Once the musclecar era began to fade in 1971, Tasca left his hot rod days behind, with no regrets, and returned to the world of luxury cars. He has alway been one to look ahead, not behind. Win on Sunday? Robert F. Tasca wins every day.

According to Tasca, "the Cobra Jet began the era of Ford's performance supremacy. In my opinion, it was the fastest production car in the world at that point. And I'm not talking top speed, I'm talking fun fast, get up and go." After witnessing a pre-production Cobra Jet run the quarter-mile in a sensational 13.56 seconds, *Hot Rod's* ever-present Dahlquist couldn't have agreed more, claiming "the CJ will be the utter delight of every Ford lover and the bane of all the rest because, quite frankly, it is probably the fastest regular production sedan ever built."

Mustang performance heated up even more in 1969, the year for yet another restyled ponycar body. That new look was certainly hot, as were a handful of new models, muscular Mustangs that appeared thanks mostly to the earlier arrival of Semon E. "Bunkie" Knudsen. On February 6, 1968, Henry Ford II had shocked Detroit—and Lee Iacocca—by hiring Knudsen, GM's former executive vice president, as Ford's president. Knudsen loved performance, and under his direction Mustang performance flourished.

Bunkie's logic was simple. In his opinion, the Mustang was "a good-looking automobile, but there are a tremendous number of people out there who want good-looking automobiles with performance. If a car looks like it's going fast and doesn't go fast, people get turned off. You should give the sports-minded fellow the opportunity to buy a high-performance automobile." That he did. Three new high-performance Mustangs debuted during Knudsen's brief 18-month stay atop Ford, the Mach 1, Boss 302 and Boss 429.

At the tail, the California Special was adorned with the Shelby's fiberglass ducktail decklid and sequential turn-signal Thunderbird taillights. Although the letters "GT" were used as part of the California Special's name, the cars weren't GTs unless equipped with that options package, which of course added, among other things, quad exhaust tips and a set of 14x6 styled-steel wheels. Total California Special production, including the 300 similar "High Country Specials" sold out of Denver, was 4,325 cars.

It's called a "Shaker," *and for good reason. The optional ram-air equipment introduced for the Mustang in 1969 protruded right through the hood, vibrating for all to see whenever a 428 Cobra Jet's pedal went to the metal. A '69 CJ Mach 1 became the "quickest four-place production car" ever tested by* Car Life, *running the quarter-mile in 13.9 seconds.*

The most civilized of the three, at least in standard form, the '69 Mach 1 "SportsRoof" (Ford's new name for its fastbacks) was many cars to many drivers, depending on the chosen underhood power source. A 351cid two-barrel Windsor small-block was standard, with optional engines including a four-barrel 351, the 320hp 390 GT big-block, and proven 335hp 428 Cobra Jet, with or without that distinctive ram-air "Shaker" hood scoop.

Whatever the engine, a Mach 1 made its presence known thanks to a whole host of standard sporty features. Blacked-out hood with non-functional scoop (unless ram-air was specified). Racing-type hood pins and dual color-keyed rac-

Reportedly, Ford only built 13 428 CJ convertibles for 1969, and this is the only known example with a ram-air Cobra Jet backed by a four-speed, the latter included as part of the Drag Pack option.

Transforming a Cobra Jet into a Super Cobra Jet, the Drag Pack added an oil cooler, a Top-Loader four-speed and a 4.30:1 Traction-Lok 9-inch rearend.

ing mirrors. "Mach 1" bodyside stripes. Pop-open gas cap. Chrome styled steel wheels. "Special Handling" suspension with E70 rubber. And that wasn't all. Inside, high-back buckets, a sport steering wheel, console, and simulated woodgrain appointments were also made part of the deal; a package that so impressed *Car Life's* editors, they named the new Mach 1 their "Best Ponycar" for 1969.

Mustang buyers were also impressed, or at least the 72,458 who chose '69 Mach 1s were. Another 40,970 customers bought Mach 1s the following year, when the equally impressive 351 Cleveland small-block joined the Mustang options list.

More purposeful and less likely to appeal to the average ponycar buyer, Bunkie's two Boss Mustangs stood at opposite ends of the performance spectrum. One was a high-winding small-block screamer, slung low to handle like nobody's business. The other was a nose-heavy, cantankerous big-block beast, built to run in a straight line, preferably a line measuring exactly one quarter mile.

Ford's Boss 302 was created in response to Knudsen's demand for "absolutely the best-handling street car available on the American market," an appropriate goal considering the plan was to take the car racing on the Sports Car Club of America's Trans Am circuit. Chassis engineer Matt Donner responded with a lowered, heavy-duty suspension and fat F60 tires on seven-inch-wide wheels. Designer Larry Shinoda, another GM refugee who followed Knudsen to Ford, supplied the spoilers, slats, and splashy graphics. And power came from an exclusive 290hp 302cid small-block topped by Ford's excellent canted-valve Cleveland heads. Brute force, however, wasn't the idea here. Wouldn't you know it, according to *Car and Driver* the Boss 302 Mustang was "the best handling Ford ever to come out of Dearborn and may just be the new standard by which everything from Detroit must be judged." Boss 302 production was 1,934 for 1969, another 6,318 for 1970.

Brute force, on the other hand, *was* the Boss 429's forte. Also unique to its application, the 375hp "semi-hemi" Boss 429 big-block was actually created to make the

Super Cobra Jets were built with hard use in mind, and were thus beefed on the bottom end to keep the crank and rods where they belonged. Also assisting in holding this hell-raiser together was this external oil cooler, mounted just ahead of the radiator on the driver's side.

Boss 302 Mustangs, built in both 1969 and '70, simply can't be missed on the street, thanks in part to their low-slung stance, those various radiant paint schemes, and that reflective bodyside stripe. This Grabber Blue '70 Boss 302 was one of 6,318 built. The Magnum 500 wheels, rear spoiler, and window slats were optional equipment.

Right: Although road handling was the Boss 302's forte, it was also no slouch under the hood, where 290 horses supplied by a canted-valve, Cleveland-head 302 small-block were ready, willing, and able to wind out all day long. But just in case a driver did a little too much winding, a 6150-rpm rev limiter (barely visible in silver just to the right and below the power steering pump's dipstick) was included as part of the package. The ram-air Shaker scoop was a Boss 302 option in 1970.

rounds on NASCAR's stock car racing circuit behind the extended snout of Ford's '69 Talladega but had to be homologated in a regular-production model before it could legally compete. NASCAR rules only asked that 500 street-worthy examples of an eligible racing engine be built; they didn't specify the particular model those engines had to go into. So the Boss 429 V8 was homologated for the NASCAR-spec Talladegas by being offered as an option for '69 Mustangs.

Clearly a race engine not intended for the street, the big, brawny Boss 429 with its huge ports wasn't exactly suited for everyday operation. Even though it could propel a Mustang to mid-13-second bursts through the quarter-mile, that performance was still below potential. In Shinoda's words, the semi-hemi 429 "was kind of a slug in the Mustang." Slug or not, Boss 429s attracted 859 buyers in 1969, followed by 499 more before production ceased in January 1970.

Almost lost in the two Boss Mustangs' shadows, the once-prominent GT made its final appearance in 1969. Production was only 5,396, the lowest total during

Autolite
BOSS 302

Hands down, the most beastly Mustang ever was the Boss 429, a hunkered-down, hairy animal with a heavy-duty lowered suspension, big 15-inch Magnum 500 wheels and 375 horses worth of "semi-hemi" V8 beneath a huge hood scoop. Scoops on '69 Boss 429s were painted to match the body, while '70 scoops were black. This Grabber Blue '70 Boss 429 was among 499 built. Another 859 had rolled down the Kar Kraft production line in Brighton, Michigan, the year before.

the five-year run. And while the GT was on its way out in 1969, the upscale Grande was making its first appearance. Offered in hardtop form only, the Mustang Grande featured wire wheelcovers, a deluxe interior, and various insulation and suspension upgrades to guarantee a quieter ride. A standard vinyl landau roof and houndstooth cloth seats were added to the package for 1970. Grande hardtop sales hit 22,182 in 1969, 13,581 the following year.

A luxury ponycar? Perhaps the times they were a-changin'. Again.

"The Boss 429 started out as a NASCAR engine. We were going to build enough to make it legal for NASCAR. In order to make it racing legal, we had to sell them. The bottom end was similar to the Super Cobra jet 429. The Boss 429 Hemi-head was really patterned after the Chrysler Hemi, to tell you the truth. The hemispherical combustion chamber still is the best from a power standpoint and surface-to-volume ratio."

—Hank Lennox, Ford Engineer

Interestingly, the Boss 429 V8 was made a Mustang option only to homologate the engine as a NASCAR-legal racing powerplant for Ford's aerodynamic long-nose Talladega. NASCAR Talladegas had Boss 429s behind their odd-looking snouts; street Talladegas used 428 Cobra Jets. Fitting the big 375hp semi-hemi V8 into a Mustang engine bay required moving the shock towers one inch farther apart. Ram-air equipment and a 735cfm Holley four-barrel were standard.

P-51

What's In A Name?

As Ford's famous galloping logo implies, a Mustang is a horse, a "wild horse of the North American plains" to be exact per most dictionaries. But while making the obvious connection between pony and car is simple enough, that's not exactly how Iacocca's baby was originally christened, at least not according to executive stylist John Najjar. Although the name did indeed initially come right off a North American plane, that was plane as in airplane. And North American as in North American Aviation, Inc., builders of the legendary P-51 Mustang fighter of World War II.

It seems Najjar had a thing for the awesome P-51, among the world's best performing fighter aircraft before the arrival of jet technology made prop power obsolete immediately following the war's end in 1945. Even then, Mustangs were ready, willing and able to take to the air in anger again five years later when hostilities broke out in Korea, then continued flying for various foreign air forces, seeing additional action as late as 1956 for Israel against Egypt. Over the years since, Mustangs have remained "in action" as airshow stars and dominating air-racers.

In all, some 15,300 P-51s were built beginning in 1941 at North American's plants in Inglewood, California, and Dallas, Texas. Initially underpowered and ceiling restricted with an unsupercharged 1200-horsepower Allison V12 driving its huge four-bladed prop, the P-51 didn't really take off until Britian's Royal Air Force officials in 1942 exchanged the Allison for their Spitfire

Some 25 years before Ford's Mustang arrived, North American Aviation was beginning production of this Mustang, the powerful P-51 fighter plane of World War II fame. This rare P-51D, owned by Henry "Butch" Schroeder of Danville, Illinois, is the F-6D photo reconnaissance derivative, a Mustang that carried both cameras and guns.

fighter's Rolls-Royce Merlin V12. Manufactured under license in this country by Packard, the superb supercharged 1400-hp Merlin transformed the Mustang from an average warplane into a dominating high-altitude hunter-killer. Top speed for the Merlin-powered P-51B was 440 mph, compared to 390 mph for the Allison-equipped P-51A. Additional modifications later upped the top-end ante to 487 mph for the P-51H.

Most plentiful among the various Mustang models was the P-51D, the first P-51 to use a modern-looking bubble canopy. Of some 8000 P-51Ds produced, 136 were modified to carry cameras as well as guns and redesignated for photo reconnaissance duty as F-6Ds. Owned by Henry "Butch" Schroeder of Danville, Illinois, the F-6D show here was painstakingly restored by Schroeder's ace righthand man, Mike Vadeboncouer, to represent "Lil' Margaret," the recon Mustang flown over Europe by Captain Clyde East. In typical fashion for a married fighter pilot, Captain East named his trusty mount after his wife.

Since, unlike most recon aircraft, camera-equipped Mustangs were heavily armed, F-6D pilots could ably "defend themselves," which Captain East did regularly. Flying for the 9th Air Force, 10th Photographic Reconnaissance Group, 15th Tactical Reconnaissance Squadron, Captain East scored 13 confirmed aerial kills, making him the leading American recon ace of the European theater. As for how well his Third Reich snapshots came out, no one quite recalls.

Whether supplying invaluable intelligence photography, defending B-17 bombers over Berlin, escorting B-29 Super Fortresses to Tokyo, strafing trains in Korea, or thrilling crowds as Reno air-racers, P-51 Mustangs have always soared high. It's little wonder John Najjar was impressed, so much so in 1962 he borrowed North American Aviation's Mustang moniker for Ford's two-seat, mid-engined ponycar "prototype."

The rest is history.

Lovingly restored, maintained and sometimes flown by Schroeder's ace mechanic Mike Vadeboncouer, this F-6D was modelled after the one flown by Captain Clyde East over Europe. As the markings on "Lil' Margaret" attest, East scored 13 air combat kills, making him the leading American recon ace.

MUSTANG
BOSS
351
BOSS 351
MUSTANG
Radial T/A

4 1971-1973

More Mustang, Less Sales

If Lee Iacocca thought the '67 Mustang had been "a fat pig," it wouldn't have taken a rocket scientist to determine his opinion of Ford's all-new ponycar for 1971. All subjective responses to relative girth aside, the redesigned '71 Mustang was indeed a big automobile, much too big, in Iacocca's eyes, to be a ponycar. Practically everything about the machine was bigger compared to its 1970 predecessor. Wheelbase was up an inch while total length grew two. Track was widened a whopping 3in in front, 2.5 in back, making room for a two-inch increase in body width. Up as well was curb weight, by about 250 pounds, and base price, a number that surpassed $3000 (for a V8 coupe) for the first time in Mustang history.

As in 1967, enlarging the Mustang in 1971 was basically a direct response to increasing competition in Detroit's hot-and-heavy horsepower race. Keeping up with the automaking Joneses in the late '60s meant producing more and more horses, which in turn meant bigger engines, which in turn meant bigger engine compartments and beefier platforms to handle all that power.

In 1970, Ford's all-new 429-cid 385-series V8—introduced for Thunderbirds in 1968—had superseded the aging FE-series V8 as the big-block performance leader in the company's full-sized ranks,

The last great Boss Mustang was the Boss 351, a one-hit wonder for 1971. Like the Boss 302 small-block, the Boss 351's 351 HO V8 relied on Ford's excellent Cleveland heads with their canted valves and excellent breathing characteristics. Supporting cast included a Hurst-shifted four-speed, 3.91:1 Traction-Lok rearend, F60x15 rubber, competition suspension with staggered rear shocks, and power front discs.

The 330hp 351 HO, for High Output, featured 11.7:1 compression, a solid-lifter cam and a 715-cfm four-barrel carburetor force fed through a ram-air hood. According to Motor Trend, *this combo beneath a '71 Boss 351 Mustang's hood was capable of 13.8-second blasts down the quarter-mile.*

leaving the vaunted 428 Cobra Jet FE to carry on for one last year as the Mustang's meat-and-potatoes muscle mill. With the newly expanded ponycar body on the scene in 1971, the bigger, better 429 CJ then replaced its 428 forerunner under that long, long hood, making for the largest, most powerful Mustang ever. Whether that result was good or bad is your call.

As for Iacocca, he of course wasn't pleased, and also wasn't shy about blaming one man for what he felt was the Mustang's wrong turn. Bunkie Knudsen, the former GM executive Henry Ford II had put ahead of Iacocca into Ford's president chair in February 1968, had needed only one look at chief designer Gary Halderman's mockup that same month to okay the basic design for the '71 Mustang. "We had never had approvals like that before," recalled Halderman, who himself wasn't all that excited about the Mustang's newfound mass. But two months after Bunkie's decision on the '71 model, the brawny 428 Cobra Jet debuted, and the Mustang was then off and running

Although the '71 Mach 1 was certainly sporty and definitely powerful, there was one negative aspect of the car most witnesses couldn't help but point out. Calling the '71 SportsRoof with its 14-degree "fastback" a "flat back," Car Life's *critics explained that the car's "rear window would make a good skylight." "A glance in the rearview mirror," they continued, "provides an excellent view of the interior with a small band of road visible near the top of the mirror." And according to* Sports Car Graphic, *"the flat, minuscule rear backlite gives a beautiful view of the stars for two in the rear, but little else." In conclusion, however,* Sports Car Graphic *concluded that "whatever (the car) isn't, it is exciting, and... no Mach 1 is going to rust in a showroom."*

towards its future as a high-powered heavyweight.

"Performance, performance, performance!" exclaimed Iacocca later, "that was what triggered [the Mustang] out of the small-car world. In 1968, in his autobiography, Knudsen added a monster engine with double the horsepower. To

The strongest Mustang powerplant in 1971 was the 385-series 429 Cobra Jet, superseding the FE-series 428 CJ of 1970. With or without optional ram-air equipment, the 429 CJ rated at 370 horsepower, enough oomph to propel a '71 Mach 1 through the lights at the far end of a quarter-mile in 13.97 seconds, according to Super Stock *magazine.*

support the engine, he had to widen the car. By 1971, the Mustang was no longer the same car, and declining sales figures were making the point clearly." Numbers didn't lie. Mustang sales dropped nearly 22 percent to 149,678 units for 1971, reflecting a descending trend that had continued steadily since 1966 as Detroit's ponycar pie was being sliced into smaller and smaller slabs.

If there was one consolation for Iacocca in 1971 it was that Bunkie wasn't around to revel in his handiwork. On September 11, 1969, only 18 months after he'd been hired, Knudsen was let go by Henry Ford II. Why the rapid firing? "I wish I could say Bunkie got fired because he ruined the Mustang or because his ideas were all wrong," wrote Iacocca. "But the actual reason was because he used to walk into Henry's office without knocking. That's right—without knocking!"

Certainly not sorry to see Knudsen go, Iacocca also recalled the event with a smile. "The day Bunkie was fired there was a great rejoicing and much drinking of champagne. Over in public relations, someone coined a phrase that became famous throughout the company: 'Henry Ford once said that history is bunk. But today, Bunkie is history.'" Fourteen

The '72 Sprint convertible's patriotic interior featured vinyl bolstered seats with cloth inserts done in red, white and blue

Based on the third of three Sprint packages offered over the years for the first-generation Mustang (following the 1966 and 1968 versions), the '72 "Olympic Sprint" convertible was a special package put together by the Washington D.C. area Ford dealers to commemorate both the 1972 Olympics and Washington's annual Cherry Blossom Parade. Only 50 of these convertibles were built, all equipped identically, beginning with the optional Sprint package available on '72 hardtops and SportsRoofs. Special red/blue striping, a unique interior and "USA" shields on each rear quarter were the most noticeable Sprint features. All Sprint convertibles also had 302 V8s, AM/FM stereo and air conditioning.

Bottom: The last of the first-generation Mustangs, the 1973 model, differed little from its 1971 and '72 predecessors, especially when equipped with the optional exterior decor group, which, among other things, added a blacked-out grille and color-keyed bumper. Different grille arrangements helped set apart the base models in 1971, '72, and '73.

Right: The attractive forged-aluminum sport wheels were a new option for 1973. Additional extra-cost features on this '73 convertible include the Gold Glow metallic paint, black-accented "NASA" hood with ram-air, and trunk-mounted luggage rack.

MUSTANG
GOODYEAR

months later, Iacocca became the top man in the Ford pecking order and immediately began promoting an idea first brought up as early as November 1969, a proposal to return the Mustang to its more polite, petite roots.

Iacocca's plan, however, wouldn't become reality until 1974. In the meantime, Bunkie's really big Mustang made its dealer debut in September 1970, then rolled on in similar fashion through three model runs. Following the disappointing 22 percent sales drop for the new '71 model, Mustang production fell again in 1972, then rebounded slightly in 1973. By that time, high performance—Knudsen's inspiration for fattening up Iacocca's pony—had ironically all but faded away from the Detroit scene as tightened emissions restrictions, stricter safety standards, and sky-high insurance rates had put the brakes on the horsepower race.

Mustang convertible buyers in 1973 automatically received power front discs, knitted vinyl buckets, color-keyed carpeting, and a power top with backlite for their $3189 base price. Helping kick up the bottom line in a hurry here are air conditioning, AM/FM stereo, power steering, console, full instrumentation, and tinted glass.

Fading away as well were Ford's muscle Mustangs, cars like the formidable Boss 302 and Boss 429, both cancelled after 1970, leaving the Mach 1 to carry on as Dearborn's highest profile ponycar. The undeniably hot 429 Cobra Jet did remain a powerful option for one more year, then it too was cancelled after only 1,865 CJ Mach 1s were built for 1971. Top performance Mach 1s in 1972 and '73 relied on the 351 Cleveland small-block, a potent powerplant that nonetheless still fell short of the brutish big-block Cobra Jets as far as true tire-melting performance was concerned. At least in basic form. A beefed-up version of the 351 Cleveland small-block capable of doing battle with most big-blocks bullies did appear in 1971 to help remind the ponycar performance faithful that it is never really over until it's over.

Introduced in November 1970, the last of the great muscle Mustangs, the Boss 351, like the Mach 1, was offered only in SportsRoof form. Beneath its blacked-out, ram-air "NASA" hood was the 330-horse "351 HO" with its free-breathing, canted-valve Cleveland heads. Serious 351 HO features included a 715-cfm four-barrel, 11.7:1 compression, and a solid-lifter cam. Power front discs, a competition suspension with staggered rear shocks and F60x15 rubber, a Hurst-shifted four-speed, full instrumentation, twin twist locks on that black hood, a chin spoiler, and honeycomb grille completed the package, an impressive one to say the least.

According to *Car and Driver*, the Boss 351 offered "dragstrip performance that most cars with 100 cubic inches more displacement will envy." In a *Motor Trend* road test, a '71 Boss 351 scorched the quarter-mile in 13.8 seconds, results comparable to a '71 Mach 1 armed with a 429 Cobra Jet. That performance also ranked the Boss 351 right up among the hottest Fords ever. Nothing like going out with a bang. Although talk of a second-edition Boss 351 did make the rounds, the car ended up a one-hit wonder as the axe finally fell on Ford factory performance in 1972.

Innocent bystanders might've noticed the year before that something was up as the '70 Mach 1's base 351 was traded for a much more timid 302 Windsor standard small-block for the '71 Mach 1. Also indicative of changing times was the Mustang Grande's popularity rise. As performance dwindled, luxury became a more prominent ponycar selling point, with Grande sales jumping 28 percent in 1971. Two years later, the vinyl-roofed Grande hardtops made up nearly 20 percent—25,274 cars—of total Mustang production for 1973, the last year for Ford's first-generation pony.

An obvious sign that performance was no longer the going thing around Detroit, optional ram-air was only offered for the Mustang's 351 Cleveland 2V in 1973, leaving buyers of the more powerful 266hp four-barrel Cleveland small-block in a huff. Net output rating for this two-barrel 351, with or without ram-air, was 177 hp.

5.0
Illinois
HOOFBT 4

5 1974-present

The Rest of the Ponycar Tale

It was. It is. Or at least so claims Ford's recent advertising campaign touting its all-new fourth-generation Mustang, an excellent upgrade of the long-hood/short-deck theme combining thoroughly modern engineering improvements with a retrofit essence of Iacocca's original ponycar flair. Replacing the long-running Fox chassis, first used way back in 1979, the redesigned SN-95 platform appeared in December 1993 sporting a newfound firmer stance and wearing a restyled skin that left many Mustangers wondering why it took Dearborn so long to leave the aging third generation behind.

A convertible returned to the Mustang lineup in 1983 for the first time in 10 years while the GT's High Output (HO) 5.0L V8, in its second year, jumped up to 175 horsepower, thanks to the addition of a Holley four-barrel carburetor. Attractive aluminum wheels on the GT hatchback in back were part of the optional TRX suspension package. T-tops were also an option.

Of course, there were some who were sorry to see the Fox-chassis "Five-Oh" Mustang go, if only because it was the car responsible for replanting the Blue Oval performance banner smack-dab in the middle of the Camaro's roof during the mid-'80s. In doing so, Ford managed to pass Chevy's reigning champion in the "best bang for the bucks" derby, holding the lead firmly until the Bow-Tie boys unveiled their own all-new long-hood/short-deck creation late in 1992. Going on three years later, Mustang fans are still waiting to see whether or not Ford will retaliate soon with a more powerful ponycar, one better suited to run neck-and-neck with Chevy's 275hp LT1 Camaro. As they say, however, time alone will tell.

Lee Iacocca's "little jewel," the Mustang II, initially impressed everyone, including Motor Trend's *staff, who named it Car of the Year for 1974. But after sales skyrocketed to 385,993 for the first of the second-generation ponycars, the gleam quickly wore off this really small gem, and not even a revival of the Cobra imagery or Mach 1 name could stem the tide. By 1979, it was time for a third-generation Mustang. This '75 Mustang II was one of 21,062 Mach 1 hatchbacks built that year.*

Long respected for the way it could put a Camaro owner in his place, the Fox-chassis Mustang was also responsible for helping revive Ford's ponycar bloodline immediately following the Mustang II years of 1974-78. Both a response to purists' complaints concerning the "fattened-up" Mustangs of 1971-73 and an attempt to adapt to changing attitudes inspired by rising fuel prices, Iacocca's "little jewel" was initially well received as what Ford ads called "the right car at the right time." *Motor Trend's* editors apparently agreed wholeheartedly, bestowing the '74 Mustang II with their coveted "Car of the Year" award.

Created about the time disco was sweeping the land, Ford's '78 King Cobra was either the right car for the times or a really gaudy representation of just how off-the-wall the tape-and-stripe guys could get. Along with a large "Cobra" hood decal, standard King Cobra features also included various striping, spoilers, and lower body spats. Suspension upgrades and eye-catching aluminum wheels were also part of the package. About 5000 were built.

In *Motor Trend's* words, the all-new second-generation ponycar was "a total departure from the fat old horse of the recent past... a rebirth of the Mustang of 1964-65—smaller and even more lithe in feel than the original pacesetter." Based on a Pinto platform and sporting sheetmetal originally designed by Italy's famed Ghia design studio, the Mustang II measured 13 inches less hub-to-hub than its 1973 predecessor and was also four inches skinnier and roughly 300 pounds lighter. New engineering features included rack-and-pinion steering, improved noise and vibration reduction, and a penny-pinching 2.3-liter four-cylinder engine standard. No V8 was initially offered, only an optional 2.8-liter V6.

Yet even without any real performance appeal, the '74 Mustang II brought buyers running into Ford dealerships. At least at first. Mustang sales nearly tripled for 1974, hitting 385,993, reminding some of the mad rush of April 1964. Then came rapid decline. A 51 percent drop in sales for 1975 was followed by a steady slide in 1976 and '77. Introduction of a lukewarm, optional 302 V8 in 1975 and the token revival of the Cobra name—appearing as the taped-and-striped Cobra II—the following year did little to stem the tide as customers grew less enchanted with the Mustang II's cramped quarters and weak performance. Although sales did rise nearly 25 percent for 1978, by then the deal was already done, and not even the high-profile King Cobra, with its flashy decals and snazzy spats and spoilers, could save the day for the second-generation Mustang.

Commonly taken out of context, the Mustang II is an easy target today for probably too many slings and arrows. But in all fairness, even those among the

In 1964, Ford's first Mustang was honored by its selection to pace the Indianapolis 500. Fifteen years later, the first third-generation ponycar received the same honor, only this time Dearborn took advantage of the situation and marketed about 11,000 Indy Pace Car replicas. Both 302 V8 and turbocharged four-cylinder versions were built.

ponycar crowd who still have a soft spot in their hearts for the little Mustang II have to admit it was one of Ford's better ideas that deserved to run its course. Quickly.

So it was that big plans for yet another all-new Mustang were in the works even as the first Mustang IIs were rolling off the ary 1973. Dearborn's first Fox-chassis cars, Ford's Fairmont and Mercury's Zephyr, debuted in August 1977, followed by a Mustang/Capri version a year later.

Seemingly defying physical laws, the '79 "Fox" Mustang was wider, longer and taller than the Mustang II, yet was 200 pounds lighter. Jack Telnack's styling team

Ford celebrated 20 years of ponycar history with a special '84-1/2 anniversary model, available as a hatchback or convertible. All 20th anniversary cars were painted Oxford White with "GT 350" identification. Power came from either a 5.0L V8 or the 2.3L turbo four. This V8 20th anniversary Mustang was one of 3,900 hatchbacks sold; another 5,260 convertibles were also produced.

trucks in rapid fashion. In December 1974, Iacocca and Henry Ford II gave two thumbs up to the Fox project, a design for a pair of fuel-efficient model lines first seen in official Ford paperwork in Febru- did the attractive shell, which rolled on a new chassis featuring MacPherson struts up front and a coil-spring, four-link suspension in back. Again, a 2.3-liter four-cylinder was standard, but optional under-

hood fare included a turbocharged four and a 130hp 5.0-liter (302 cid) V8. Signs perhaps that performance and prestige would again become Ford Mustang's middle name? Yes, according to *Motor Trend's* John Ethridge, who claimed the third-generation pony could "now compete both in the marketplace and on the road with lots of cars that used to outclass it."

As in 1974, Mustang sales again soared, jumping some 92 percent in 1979 to 369,936 cars. And as in 1964, the new Mustang was chosen to pace the Indianapolis 500. About 11,000 regular-production Indy Pace Car replicas were built with both 5.0L and turbo four power. Included in the deal were special graphics and the sport-minded TRX suspension.

Suitably honored and definitely hot to trot, the Fox-chassis Mustang then set out on a steady uphill climb towards Detroit's low-priced performance pinnacle. First came the 157hp High-Output (HO) 5.0L two-barrel in 1982, the year the Mustang GT returned after a 12-year hiatus. In 1983, a Holley four-barrel went atop the 175hp 5.0L HO and a convertible rejoined the ponycar lineup for the first time since 1973.

Left: Introduced in 1984, Ford's SVO (Special Vehicle Operations) Mustang returned as '85, '85-1/2 and '86 models. All were easily identified by their assymetrical hood scoops, aero noses, and multi-level rear spoilers, but the most intriguing aspects of the SVO involved its turbo-four power source and its hot-handling suspension. In all, 9,844 were built, including 3,314 '86 models, like this one, for domestic sales.

Initially rated at 175 horsepower, the SVO 2.3-liter intercooled turbocharged four-cylinder jumped to 205 horses in 1985, then dropped back to 200 horsepower for the car's final year in 1986.

Big news for 1984 involved the debut of the Euro-style SVO Mustang, initially powered by a 175hp intercooled 2.3-liter turbo four. Featuring Koni gas-charged shocks, four-wheel discs, 16-inch wheels, and a Hurst-shifted five-speed, the SVO Mustang hit 205 horsepower in 1985, then was detuned down to 200 in its final form for 1986. In all, 9,844 were built during the three-year SVO run.

Aero styling up front was added in 1987, as were 15-inch "turbine" wheels for the GT. Essentially unchanged, the same package carried over into the following year, as this '88 GT convertible attests. Restyled 16-inch GT wheels were added in 1991.

As for additional basic upgrades, quad shocks were added in back for 1984, just in time to help celebrate the Mustang's 20th birthday. Offered as a hatchback and convertible with either 2.3-liter turbo four or 5.0 V8 power, the limited-edition '84-1/2 20th Anniversary models featured Oxford White paint with red "G.T. 350" lower bodyside stripes.

In 1985, tubular headers with dual exhausts (from the catalytic converters back) and roller lifters boosted HO output to 210 horsepower, the most for a carbureted 5.0 small-block. True dual exhausts were added the following year, when the 5.0's four-barrel carb was exchanged for the EEC (electronic energy control) EFI (electronic fuel injection) equipment. By 1987, the EFI 5.0 HO was putting out 225 horses, enough to propel the revised "aero-look" Mustang

Few could beat the Mustang's electronic fuel-injected (EFI) 5.0L HO in 1988. With a roller cam and tubular exhaust headers, the 5.0 HO put out 225 real horses, power enough to make the lightweight LX coupe a low-14-second screamer in the quarter-mile. Camaro owners never had a chance.

through the quarter-mile in a tad more than 14 seconds. Camaro drivers didn't stand a chance, and growing popularity of the hot 5.0 Mustang—either in GT or lighter LX form—helped convince Dearborn to give up ill-advised plans to replace the Fox-chassis Mustang with a Mazda-based front-driver, a platform that instead appeared as the '89 Probe.

Recognizing a winner, designers then basically left the Fox chassis well enough alone, with one of the most noticeable changes being larger, restyled 16-inch GT wheels, added in 1991. Ford did, howev-

A revised, greatly stiffened chassis and four-wheel disc brakes made the all-new '94 Mustang a much more firm, sure-footed ride compared to its Fox chassis predecessors.

And while GTs were all powered by the ever-present 5.0L V8, base models dumped their standard four-cylinders in favor of a 145hp 3.8L V6.

Although it was downrated to 215 horsepower, the '94 Mustang GT's 5.0L HO small-block is still a free-wheeling, fun machine. Notice the new brace added between the cowl and upper shock towers, one of the many engineering additions made to make the '94 Mustang platform more rigid.

er, save the best for last in 1993 as the Special Vehicle Team rolled out its 235hp Cobra, a certified throwback to the street racers Carroll Shelby once built. Even more reminiscent of Shelby's handiwork was SVT's R-model '93 Cobra, which, like Shelby American's GT 350R of 1965, meant business. As if you couldn't have guessed, "R" still stands for "racing."

And the Mustang still stands for fun, even after three decades on the road. "It is what it was, and more," announced ads for Ford's all-new fourth-generation

Left: Like so many other things in life, the new Mustang GT's two-tone interior has inspired both love and hate from the masses.

ponycar. Who's to argue? Consider the '94 Mustang's standard four-wheel discs and its improved chassis/body design that did away with the NVH (noise, vibration, harshness) problems inherent with the Fox chassis; throw in superior handling and a new base 3.8L V6 instead of a standard four-cylinder, and there your have it—the best Mustang yet.

Can it get any better? Ask again in another 30 years.

Venom for the '94 Cobra came from a pumped-up version of the GT's 215hp 5.0L small-block. A more aggressive cam, GT40 cast-iron heads, revised intake gear and less restrictive exhausts let loose an additional 25 more horses for the Cobra. Unfortunately, installing the larger induction setup meant losing the new Mustang's shock-tower-to-cowl brace, a piece that helped stiffen the latest ponycar platform.

Left: Differences setting a '94 SVT Cobra Mustang apart from its GT cousin include clear-lens headlights, special front fascia with round foglamps, chromed 17x8 wheels (with dark accents on Indy Pace Car convertibles—wheels on Cobra coupes are fully chromed), appropriate yet nearly unnoticable coiled-snake fender emblems, and a unique rear spoiler. Only 1000 Indy Pace Car replicas, all Rio Red '94 Cobra convertibles with saddle interiors and tan tops, were built by Ford's Special Vehicle Team. SVT also rolled out 5,009 '94 Cobra coupes in red, black and white.

6 Variations On A Theme

Ponycar Spin-Offs from Shelby to Saleen

Ford's popular ponycar has meant various things to various people during its three decades on this planet. Budget buggy with a sporty flair. Road-hugging street racer. Tire-melting Saturday night warrior. Semi-plush boulevard cruiser. The Mustang has worn all these hats and more, thanks both to the Better Idea guys in Dearborn and a few independent thinkers outside the company.

Not all Mustang variants were meant for the streets. While Carroll Shelby was busy in California building his GT 350 road warriors, the speed kings at Holman-Moody in Charlotte, North Carolina, were working on a red-hot Mustang of their own. Holman-Moody's product was an altered-wheelbase drag car intended for Factory Experimental (FX) competition. This '65 A/FX Mustang, one of 15 built by Holman-Moody, was originally modified for Larson Ford in White Plains, New York. It is also one of seven equipped with Ford's outrageous SOHC 427 V8. The other eight were originally powered by 427 "High-Riser" wedge-head V8s.

Some of these outsiders, like the legendary Carroll Shelby and Steve Saleen on the West Coast are certainly well known. Others, like Dario Orlando in Florida, are less so, but their creations speak for themselves, loud and clear. Specialty Mustangs, from Shelby's first fire-breathing GT 350 in 1965 to Saleen's latest low-slung S-351, stand as just as much a prominent part of Ford's ponycar history as the regular-production models and rightly so since Dearborn has been more than willing to "support" these various projects one way or the other over the years.

Shelby, of course, kicked things off in the specialty Mustang market not long

With an overhead cam in each head and hemispherical combustion chambers, the 427 SOHC, or "Cammer" as it was known, could wind out like no other big-block out there. Output was in the 600 horsepower neighborhood. Although various rumors beginning in 1964 spoke of a regular-production Cammer, no 427 SOHC V8s ever made it to the street in a standard-issue Ford.

after Iacocca's first pony hit the road some 30 years ago. At the time, the famed Texas-chicken-rancher-turned-race-driver was two years into production of his little Anglo-American hybrid racer, the AC Cobra, a British sports car shell powered by a warmed-over Blue-Oval small-block V8. Dearborn officials, however, preferred to see a more recognizable Ford product taking checkered flags, and what better candidate for that job than the new Mustang? A deal was then made to supply the Shelby American plant in Venice, California, with bare-bones Mustang fastbacks and Shelby's crew did the rest, with production of the race-ready GT 350 beginning in October 1964, followed by the official public unveiling January 27, 1965.

"Pretty much a brute of a car," according to *Road & Track*, the '65 GT 350 was meant for one thing, and a night on the town wasn't it. As Shelby later told *Sports Cars of the World* magazine in 1971, "my idea was to build a car that would outrun the Corvette and Ferrari production cars." Winning Sports Car Club of America (SCCA) B/Production races was the goal, and Shelby's GT 350 didn't disappoint.

"The thing was a racing car in disguise and not much of a disguise at that," recalled *Car Life's* Joe Scalzo in 1969. There were no ifs, ands, or buts about the first GT 350 and no nonsense, either. No backseat, no automatic transmission, no color choices—all '65 GT 350s had black interiors and white exteriors with optional Guardsman Blue "LeMans" stripes available. Beneath that pale skin was a bone-jarring suspension featuring lowered upper A-arms and special underhood bracing up front, over-ride torque control arms in back, and Koni adjustable shocks all around. A bullet-proof racheting Detroit Locker rearend was standard, as were the loud cutout exhaust pipes exiting directly in front of each rear tire.

Power for the beast came from a modified Hi-Po 289 fed by a 715-cfm Holley four-barrel on an aluminum intake. An enlarged aluminum oil pan, special "Tri-Y" headers and twin glass-

machine that, in the opinion of *Car Life's* Jim Wright, was what "most of us wanted the original Mustang to be in the first place." Shelby American built 562 GT 350s for 1965, including 37 truly serious R-model all-out racers.

Although Shelby American tried the trick again in 1966, pressures from Dear-

Per Ford's instructions, Shelby American toned down its GT 350 image in 1966, adding full exhaust pipes, a backseat, an optional automatic transmission and multiple paint choices. Rear quarter glass and bodyside scoops were also added.

pack mufflers were also part of the deal, which added up to 306 horsepower. Throw in a scooped fiberglass hood, a four-speed, 9.5-inch front discs, and heavy-duty 15-inch station wagon wheels, and you were ready to race in a

born to tone down the GT 350's roughness left Carroll Shelby with a bad taste in his mouth. Ford wanted a wider market for the car, with the obvious goal being to sell more Shelby Mustangs to drivers who didn't necessarily want to go racing.

Although changes were made outside and underneath, the GT 350 power source in 1966 was still the "Shelby-ized" High Performance 289, rated at 306 horsepower in Shelby trim.

Shelby updates in 1967 included various fiberglass add-ons up front and at the tail and a new model, the big-block GT 500, powered by a 355hp 428. This '67 GT 500 has the optional Le Mans stripes.

Previously standard, the Koni shocks and Detroit Locker went to the options list, the rear torque arms were replaced by simpler underslung traction bars, and the lowered front suspension was discontinued to cut costs. Full dual exhausts went underneath, a backseat went inside, and an optional C4 automatic joined the standard Borg-Warner T-10 four-speed. Five paint choices were also added, as were functional bodyside side scoops and distinctive rear quarter

Shelby GT 350 and GT 500 production was transferred from the Shelby American works in California to the A.O. Smith facility near Detroit in 1968. Then in 1969, an even more stylized Shelby Mustang appeared featuring a radically redesigned nose. Tail features remained similar to the '67-68 look with its sequential turn signal taillights and ducktail spoiler. Center-mounted exhaust tips were new for 1969. Standard power for this Black Jade '69 GT 350 was a 351 Windsor V8.

glass. Clearly more socially acceptable, the '66 GT 350 attracted nearly 2400 buyers, including the Hertz Rent-A-Car company, which put 1000 Shelbys into its fleet in 1966.

Despite the sales success, Shelby never did like the direction his specialty Mustangs took after 1965. Today, his favorite is still that first GT 350, "a no compromise car [that was] built to get the job done." But that changed once Ford began taking more interest in the project in 1966. "All of the corporate vultures jumped on the thing and that's when it started going to hell," recalled Shelby in 1971. "I started trying to get out of the deal in 1967, and it took me until 1970 to get production shut down."

Of course, Shelby's standards were much higher than the typical enthusiast,

Regardless of what engine was under the hood, Shelby Mustang interiors were always sporty and purposeful. Notice the shoulder harnesses inside this '69 GT 350.

Convertible Shelby Mustangs first appeared in 1968, the same year the GT 500KR—"KR" for "King of the Road"—debuted. Heart of the KR was the new 335hp 428 Cobra Jet V8. The KR designation was dropped in 1969 as all GT 500s were Cobra Jets, as is this big-block Shelby convertible, one of 335 built during the 1969-70 run.

who continued buying Shelby Mustangs even as the car's creator was trying to put on the brakes. Sure, Shelbys to follow may not have been as mean as the '65, but they did have their own appeal, both from an image perspective and as real performers.

Shelby news for 1967 included revised styling and a new model, the big-block-powered GT 500. An extended fiberglass snout and Kamm-back tail improved Shelby Mustang looks considerably in 1967, and an even more distinctive facade was created in 1969 once Ford stylists took complete control. Following the 1967 remake, Shelby Mustang production were transferred from Shelby's Los Angeles plant to the A.O. Smith company in Livonia, Michigan, where the renamed "Shelby Cobra GT 350/500" was built up through 1970.

A convertible Shelby appeared in 1968, as did the midyear GT 500KR, "KR" standing for "King of the Road." Heart of the GT 500KR was the 335hp

With the project all but winding down, Shelby production for 1970 basically involved repackaging 1969 leftovers. "Repackaging" meant adding black stripes to the hood and a chin spoiler below the nose. This '70 GT 500 is powered by a 428 Super Cobra Jet, an engine created by adding the Drag Pack option, which among other things included an easily identifiable engine oil cooler in front of the radiator.

428 Cobra Jet, a musclebound big-block that revived memories of the bully the Shelby Mustang had once been. By 1969, however, the end of the trail was in sight, and only remarketing leftovers into 1970 kept the Shelby Mustang alive for one last fling.

Fourteen years later, SCCA racer Steve Saleen picked up the ball and joined the specialty Mustang game, forming Saleen Autosport in Petulma, California. Initially, the Saleen Mustang's main attraction was its "Racecraft Suspension" package, which lowered

Like the almost-forgotten Mustang I racer of 1962, the ASC/McLaren Mustang convertibles built between 1987 and 1990 were all two-seaters. One of a mere 65 built for 1990, this ASC/McLaren Mustang is the only one featuring a silver exterior and maroon top and interior. Although the folding tops on these cars were manual, they were easily stowed, even by one person.

the car about 1.5 inches and added, among other things, Bilstein shocks and struts and Goodyear Eagle rubber on 15x7 Hayashi alloy wheels. Sure, the '84 Saleen Mustang was also a real looker with its distinctive striping and functional airdam, spoilers, and skirts, but it was the car's superb track-ready handling that made its $14,300 sticker seem a steal.

Up through 1993, Saleen Autosport had delivered more than 2600 of its hot-handling horses, all sporting that unmistakable Saleen rear spoiler. Advancements over the years included adding 16-inch wheels in 1986, an upper shock tower brace and four-wheel disc brakes in 1987, and a new model in 1989 to help celebrate the Mustang's 25th birthday.

Not an official 25th anniversary Mustang, the '89 Saleen SSC was nonetheless a celebration in itself, with a specially modified 300hp 5.0L beneath its hood and a whole host of heavy-duty hardware throughout. Included as well, of course, were the proven Racecraft parts, high-profile Saleen body add-ons, and various special SSC graphics and interior appointments. Priced at $36,500, the limited-edition '89 SSC hatchback attracted only 160 buyers. Even more limited was the equally impressive 304hp Saleen SC introduced in 1990. A mere 11 were built

Although cash-flow difficulties have severely limited total Saleen production after 1990, the intriguing breed has managed to survive. And with the arrival of the fourth-generation Mustang came Steve Saleen's latest project, the S-351, with its 351 Windsor small-block, an engine that reportedly dynoes at around 370 horses. Whatever the '65 Shelby "was," this latest specialty Mustang "is" in spades. Zero to 60 in 5.9 seconds, a quarter-mile pass in 8.4 ticks more, a top speed of, say, 170 mph, and 0.97g on the skidpad, all for $33,500. Does it get any better?

Dario Orlando thinks so. Orlando's Steeda Autosports, in Pompano Beach, Florida, is one of the newest kids on the specialty Mustang block, having been in the business for some seven years now. Although Orlando prefers to concentrate on the sales of Motorsport parts and various exclusive Steeda hop-up kits for five-liter Mustangs, his shop had converted, between 1988 and '93, about 400 Fox-chassis Five-Ohs into some of the best handling, hottest-running, quickest braking ponycars on the road. And they didn't stop there.

Like Saleen, Steeda Autosports has also jumped on the all-new Mustang bandwagon for 1994, introducing easily the best Steeda Mustang yet. For about $7000 above a base '94 Mustang sticker—$17,300 for the coupe, $22,000 convertible—Steeda adds its proven G-Trac sport suspension, which includes lowered, stiffer springs, urethane bushings and cross-bracing up front, an additional sway bar (joining the stock unit) in back, and 17-inch wheels at the corners. Standard Mustang four-wheel discs are upgraded with Steeda's own high-performance carbon-metallic pads, and power comes from a 290hp 5.0L with GT40 heads and intake. Orlando plans to build 100 of these '94 super 'Stangs. Not as noticeable on the street as a Saleen, a Steeda Mustang nonetheless can give its West Coast counterpart all it can handle.

Known for their excellent Racecraft Suspension, Saleen Mustangs also stand out in a crowd thanks to their large rear spoilers, 16-inch American Racing basketweave wheels (in this case), and noticeably lowered stance. Reportedly, 105 black Saleens like this one were built in 1987, with total production for all colors reaching 278.

In between the time Saleen set up shop in California and Steeda got rolling in Florida, two Livonia, Michigan, firms joined forces to create another specialty Mustang, this one concentrating more on prestige. Engine builder Bruce McLaren and the American Sunroof Company first teamed up in Livonia late in 1983, working together to produce the ASC/McLaren Capri, a distinctive Fox-chassis machine offered up through 1986 in coupe and convertible form. Founded in 1965 by West German engineer Heinz Prechter, ASC has since grown into the world's leading OEM convertible conversion company, having produced more than 400,000 droptop models for 12 different manufacturers, foreign and domestic. Today, 25 percent of convertibles sold worldwide are ASC products.

In 1987, ASC/McLaren shifted its focus to Ford Motor Company's other Fox platform, introducing an attractive two-seat Mustang roadster that showed off ASC's convertible conversion expertise. The process began with a specially prepared LX "notchback" coupe supplied off Ford's assembly line with convertible cowl support and chassis bracing. After delivery to ASC, the LX Mustang's top was sheared off, the windshield frame was bent back some 10 degrees, more reinforcement was added, and a special "tub" was dropped into the abandoned rear seat area as a housing for the folding Cambria soft top. A rear-hinged steel deck then went on to cover the top storage area. GT lower body cladding, a leather-wrapped steering wheel and shift knob, unique wheels, special taillight covers, and a new rear fascia completed the package. Options included Recaro leather seats and all other typical extra-cost Mustang features.

Production of ASC/McLaren Mustang convertibles was 479 in 1987, 1000 in 1988, 247 in 1989, and 65 in 1990, the last year for the unique two-seater, a car that proved that specialty Mustangs could be classy, too.

Whoever said variety is the spice of life must have been a Mustang owner.

Southern Californian Steve Saleen's counterpart across the country is Dario Orlando, of Steeda Autosports in Pompano Beach, Florida. Orlando's Steeda Mustangs also rely on a superb suspension, the "G-trac" system, and in the case of this '94 model, are powered by a 290hp derivative of the GT's 215hp 5.0L HO small-block. Various options and performance packages can tailor a Steeda Mustang to the individual driver's tastes.

MASH
4077
ARMY

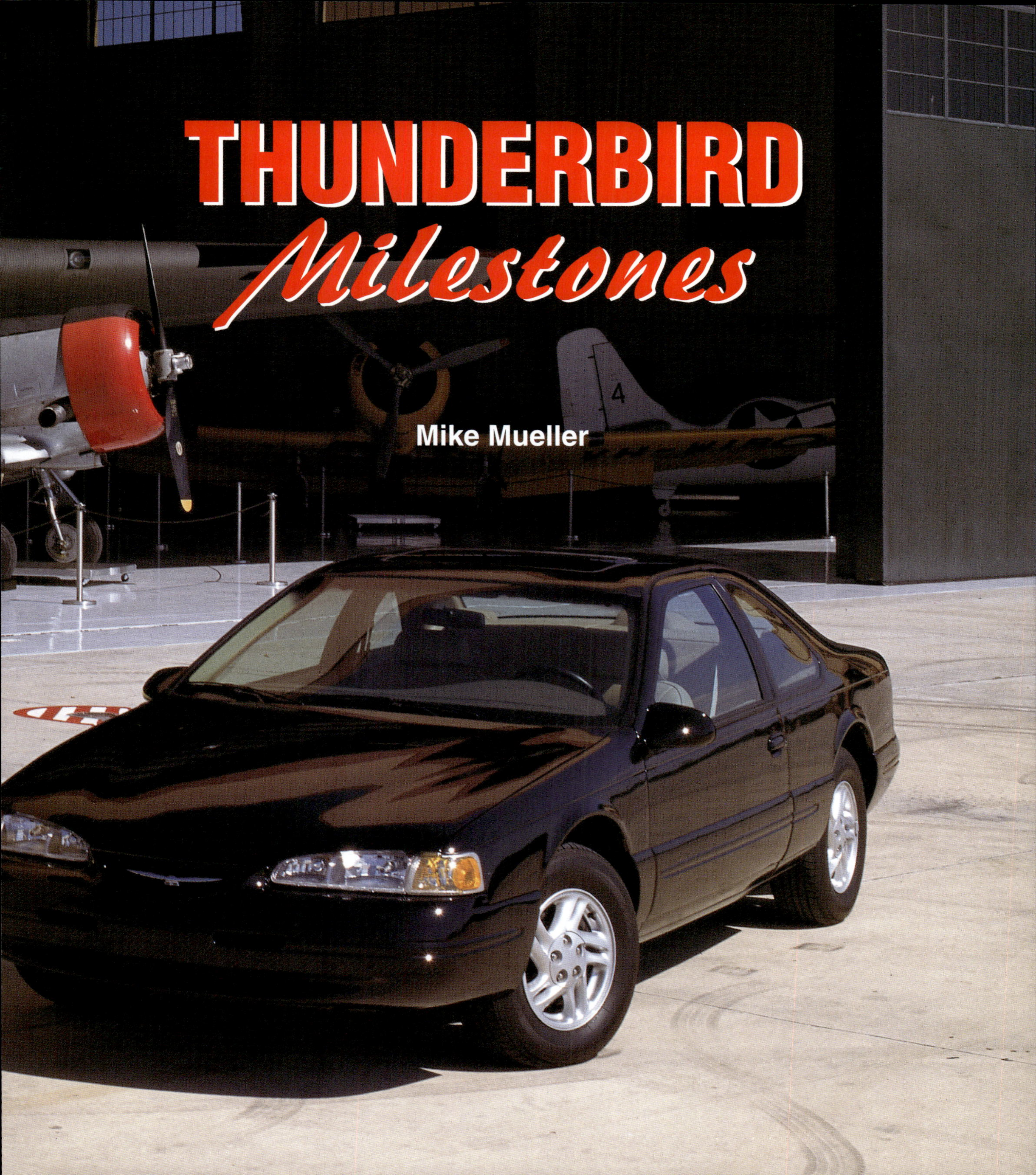
THUNDERBIRD
Milestones
Mike Mueller

Acknowledgments

Most questions concerning the T-Bird's revival will likely have been answered by the time you turn this page. What was the future for this humble scribe in 1998 is now late-breaking news for anyone within eyeshot of these words. So it is left to these 96 pages to remind you–and hopefully a few gazillion other readers–of the rich Thunderbird heritage that came before. It is lucky for all of us that so many devoted T-Bird fans have been so eager over the years to preserve that heritage. Lucky for me, too. Without the enthusiastic assistance of so many wonderfully energetic 'Bird watchers, this effort would have never taken flight.

First and foremost I must mention my good friend Max Dilley back home in Urbana, Illinois. Max and I originally crossed paths about five years back when I was working on a Mustang book, and he was still heavily involved in pony-car collecting. His easygoing enthusiasm made those first photo shoots no sweat, just fun and friendly. Max not only opened his garage for me, he opened his home. In the years since he has also never failed to answer the phone with a warm and welcoming "can do," just as he did in December 1997 when I came calling with a last-minute photo request involving his beautiful blue 1957 'Bird.

On short notice, and with less-than-desirable weather conditions making for a miserable Midwestern morning, Max came through yet again, driving his T-Bird some 15 miles despite the threat of rain. Or maybe even snow. Only after the shoot had been scheduled did I learn that his mother, Ruth Dilley, had passed away just the week before. I felt pretty small, but Max wouldn't let me feel bad, nor would he let me down. Life apparently does go on, as does my belief that I just may be one of the luckiest guys on the planet for having friends like Max Dilley.

Ray Quinlan is another friend who has also bent over backwards for me time and again whenever I've passed through Champaign, Illinois, looking for a helping hand. I would also like to thank my brother, Dave Mueller, who has always been there for me–on photo shoots or otherwise; my brother-in-law, Frank Young, who works so cheap it almost makes me feel guilty–almost; another brother, Jim Mueller Jr., and my father, Jim Sr., both also never shy about dropping everything when I come demanding. Last, but in no way least, is my very best friend, Joyce Tucker, who continually amazes me with her energy as well as her selflessness. You go, girl–and take me with you.

Glorious Thunderbird memories have been preserved for the ages thanks to enthusiasts like Max Dilley, shown here applying some TLC to his 1957 'Bird in front of his Urbana, Illinois, home.

Then there are all the club people whose friendly support was also greatly appreciated: Harvey Hodges in Champaign, Illinois; Jim Hack and George Watts of the Classic Thunderbird Club International in Signal Hill, California; Ken Leaman and Malcolm Green in New Jersey; Dr. Richard Schatz of the International Thunderbird Club in Sugarloaf, Pennsylvania; Richard and Kathy Goodroe, Elton Roberts, and Alton Maxwell, all of the Atlanta Classic Thunderbird Club.

Kudos go as well to Gil Baumgartner and Eric Larson. Baumgartner, a renowned Thunderbird restoration expert from Suisun, California, made 15 minutes of fame possible with Bo Cheadle's fabulous 1957 Battlebird. Great job, Gil, and thanks so much for your patience. Larson, curator of the Octave Chanute Aerospace Museum in Rantoul, Illinois, opened up his museum for a photo shoot featuring various representatives of the Air Force's Thunderbirds Air Demonstration Team. Once that was done, photographer Kevin Greunwald at Nellis Air Force Base in Nevada also supplied some nice shots of Thunderbird jets in action.

Additional photography came from various other equally willing sources. Bob Peterson, of Cedar Rapids, Iowa, sent along photos of his 1972 T-Bird, which just happens to be the one-millionth 'Bird built. Gene Makrancy, an incredibly energetic enthusiast and automotive historian from Port Vue, Pennsylvania, delivered an image of an experimental Squarebird done in stainless steel. Veteran automotive photographer/writer Tom Glatch FedEx'd his rendition of a 1961 Indy Pace Car 'Bird from Wauwatosa, Wisconsin. Fellow freelancer John Lee did the same with a 1963 "Princess Grace" model. Yet another welcomed photographic addition, this one featuring a pair of T-Birds from 1956 and 1962, came from cohort-in-crime Steve Statham, of Austin, Texas. Just keep rememberin' the Alamo, Steve.

Archival photos came from Dan Erickson at the Ford Photographic Library. Bob Plumer, of Drag Racing Memories (200 N. Kalmia, Highland Springs, Virginia 23075), sent along a Thunderbird Funny Car image, just one of countless historical drag racing photos offered by his company. Other racing shots came from longtime trackside photographers Dorsey Patrick, Phillip Salazar, and Bob Jackson. Donald Farr at Dobbs Publishing Group also saved my bacon with an archival picture or two. Thanks again, Donald–both for the photos and for giving me my start in this business.

Finally, many thanks go to the car owners who shared their pride and joy with my Hasselblad. In general order of appearance, they are: 1957 Thunderbird (blue), Max and Joyce Dilley, Urbana, Illinois; 1955 Thunderbird serial number 0005, George Watts, Orange, California; 1957 Thunderbird (red), Richard and Kathy Goodroe, Annistown, Georgia; 1963 Sports Roadster, Alton Maxwell, Tucker, Georgia; 1960 Thunderbird convertible, Tom Schirra, Marietta, Georgia; 1983 Thunderbird, Harvey and Adella Hodges, Champaign, Illinois; 1964 Thunderbird, Lloyd and Shirley Spellmeyer, Melvin, Illinois; 1979 Thunderbird, Vern Bruner, Rantoul, Illinois; 1956 Thunderbird, Burgess and Angela Stengl, Austin, Texas; 1962 Thunderbird convertible, George and Rosemary Duke, Austin, Texas; 1957 supercharged Thunderbird (white), Kevin and Shellie O'Hara, Orlando, Florida; 1958 Thunderbird convertible, Charlie and Diane Kidd, Palm Coast, Florida; 1963 Thunderbird Landau hardtop, 1963 Sports Roadster, and 1967 Thunderbird four-door, Ken Leaman, Fairfield, New Jersey; 1972 Thunderbird "One-Millionth," Bob Peterson, Cedar Rapids, Iowa; 1975 and 1978 Thunderbirds, Malcolm Green, Califon, New Jersey; 1988 Thunderbird, Vern Bruner, Rantoul, Illinois; 1955 Thunderbird (red), Jim and Carol Lytle, Gotha, Florida; 1997 Thunderbird (last built), Classic Thunderbird Club International, Signal Hill, California; and 1957 Battlebird, Robert "Bo" Cheadle, Pleasanton, California.

Introduction: A Feather in Ford's Cap

The day was September 4, 1997. The place was Ford's Lorain, Ohio, assembly plant. The event was a send-off for the last of a breed. After 43 model runs and 4.3 million cars, Dearborn's high-flying Thunderbird legacy came to a quiet end that evening, almost a full day ahead of schedule. Why pull the plug on one of Detroit's most popular nameplates of all time? Dwindling sales in a market that was being sliced up into smaller and smaller segments had forced Ford's hand. The decision to ground the once-proud T-Bird was nothing personal, it was business, all business. This time there would be no last-minute call from the governor, no listening to nostalgic pleas to preserve a historic legend like those that had saved the Mustang from an unpopular transformation nearly 10 years before.

"Of course we received a ton of calls when we finally made the announcement," said Ford Division Public Affairs Chief Jim Bright. "But once the decision was made, there was no turning back. Ford just let the Thunderbird run its course."

It might have been a sad moment in Lorain last September, yet nary a tear was shed. Dearborn officials did not even mark the moment with any special fanfare for the press. Again, an explanation for their actions was simple: this wasn't good-bye, only farewell for now. As this humble testament goes to press in 1998, Ford Motor Company people are busy working on an all-new Thunderbird, a car scheduled to appear with the coming of the next millennium. Rumors abound concerning plans to bring back the sporty personal-luxury car originally embodied by that classic long-hood, short-deck profile of more than four decades ago.

This is what we know. On May 8, 1998, Ford Automotive Operations President Jac Nasser announced that his company would be reviving the Thunderbird legend not all that far down the road. Nasser's announcement provided just enough information to substantiate rumors.

"A sporty Thunderbird will once again grace America's roads," he said. "Ford owes car lovers everywhere exciting, fun-to-drive vehicles. With the new T-Bird, we

Ford marked the end of a legend in 1997 by donating the very last Thunderbird built to the Classic Thunderbird Club International in Signal Hill, California. Here, the last of the breed poses with what is commonly recognized as the first, CTCI member George Watts' serial number 100005 1955 Thunderbird.

promise to bring back the magic of owning and driving an American icon." Reportedly, this next-generation 'Bird will, in Ford press release words, blend "leading edge technology and world-class driving dynamics and features with design cues true to the heritage of the original T-Bird of the mid-1950s."

According to Nasser, the reborn Thunderbird will be a rear-wheel-drive specialty sports coupe and will share a platform with the Lincoln LS, planned for 2000. Both models will be built at Ford's Wixom assembly plant in Michigan. Recreating the T-Bird ideal is right now in the hands of Chief Program Engineer Nancy Gioia and Chief Designer Doug Gaffka.

So there you have it. All else remains speculation in the summer of 1998. Best guesses put a price in the $30,000 to $35,000 ballpark. And some are predicting that Ford might even install the Mustang Cobra's 4.6-liter V-8, with its overhead cams and 305 horsepower.

All wild guesses about power sources and platform status aside, there is one thing for certain concerning the next all-new Thunderbird–it will be a great car.

This October 1954 ad in the *Saturday Evening Post* jumped the gun a bit. Initial plans to adorn the first Thunderbird with trim similar to the 1955 passenger-line style were scrapped at the last moment.

Three men and their babies—(left to right) retired stylist Alden "Gib" Gilberson, who originally named the Thunderbird; William Boyer, former vice president of design for the 1955 T-Bird; and Jack Telnack, recently retired Ford design chief. The trio met to honor the Thunderbird's 40th anniversary. *Ford Motor Company*

Dearborn's top decision-makers wouldn't have it any other way. They never had any intention of letting such a popular nameplate simply die off, even though they were willing to cut loose all old ties to the existing legend a year ahead of schedule. Squeezing the life out of the T-Bird for one last year before a total rebirth served no purpose except to drag a grand image down. Besides, interrupting the run has only helped heighten anticipation for what's coming next–the greatest Thunderbird yet, perhaps?

Of course, fanatical followers of those first two-seat classics will always make a case against that claim. Dearborn designers in 2000 will undoubtedly remind modern-day buyers of the original Thunderbird ideal, but they will probably never top the "Littlebirds" for the way they captivated the country in 1955, 1956, and 1957. Those epic four-wheel flights of fancy are still able to hypnotize today's car crowd with their stunning style, sex appeal, and timeless impressions. According to a 1992 *USA Today* poll, the 1957 Thunderbird is "America's Favorite Classic Car."

"Named after a legend, it may create one of its own," predicted a January 1955 *Motor Life* report on the newly introduced Thunderbird. Prophecy or understatement? Customers were already showing up in droves at Ford dealers as that article was being written. Orders surpassed 3,500 in just 10 days after the 1955 Thunderbird's public introduction in October 1954. Initial production plans called for

10,000 T-Birds that first year. The final tally was 16,155. After a slight dip in 1956, production of the last–and in most minds the best–two-seat T-Bird in 1957 reached 21,380. Not bad at all for such a narrowly focused automobile, a car that had overnight created its own tight little niche in the American market.

Inspired in rapid-fire fashion by Chevrolet's fiberglass plaything, the Corvette, Ford's first-edition Thunderbird wasn't exactly a sports car–something most observers quickly pointed out. It was too heavy, too luxurious, too comfortable. It had roll-up windows, a big V-8, and power options. Nonetheless, Corvette and Thunderbird were initially classed as rivals, if only because they both had two seats. Hot performance was available in each case, but serious horsepower and handling were the Corvette's realm. Ford in 1955 preferred to label the Thunderbird as a personal driving machine, a tag that soon became "personal luxury." After 1957 these two fun little two-seaters went their separate ways, never to meet again in the same marketing breath.

The Thunderbird ended up taking many different paths itself during its 40-plus years on the road. There remains today some purists who openly insist that Ford stopped building Thunderbirds after 1957. Room for two represented the only way to fly; anything else was blasphemy. Still others consider 1966 the last year. Four doors and accommodations for the entire crew didn't quite cut it for this faction.

But Ford did roll out Thunderbirds each year from 1955 to 1997. In all, 11 different generations appeared featuring many different cars for many different drivers. Luxury, or at least a modicum of such, has always been included in the deal even if the model in question was no longer so personal. The Thunderbird began small, got bigger, grew huge, shrunk a little, and then got smaller still. The sporty two-seater was replaced by a four-place luxo-cruiser, then a four-door palace on wheels, a trimmed-down showboat, followed by a downsized market-conscious machine, then an aerodynamic award-winner, and finally a thoroughly modern sports coupe. These changes reflect distinct historical differences in the car's development.

True diehards still feel Ford stopped making Thunderbirds after 1957. Still another group gave up the ghost after 1966 when four doors were added into the equation. Shown here are three definitions of pre-1967 personal luxury: a 1957 in front, backed by a 1960 (left), and a 1963 Sports Roadster.

In the beginning, the Thunderbird went through a series of "three-year plans." Two-seaters were built from 1955 to 1957, followed by the first four-place 'Birds from 1958 to 1960. Though a disappointment to the fast lane crowd, who preferred their personal luxury in a more intimate, sexier body, the second-generation "Squarebirds" represented the difference between selling like hotcakes and selling by the boatload. Ford Division Chief Robert McNamara's decision to widen the Thunderbird's scope by enlarging its parameters in 1958 was a wise move, one that kept the 'Bird flying into a new decade. With family men now able to partake in a little Thunderbird prestige, sales of the car almost doubled for 1958 and continued rising each year up through 1960.

Sales slipped a bit, yet remained strong during the "Roundbird's" three-year tenure from 1961 to 1963. Highlighting this period was the super-cool Sports Roadster, a sleek machine that revived still-warm memories of the Thunderbird's

Eleven generations have come and gone since 1955. The 1957 Littlebird in front is joined by the first "Aerobird," introduced in 1983, and a 1964 hardtop.

earliest days thanks to a tonneau cover that hid the back seat beneath twin racing-style headrests. Sports Roadsters were offered in 1962 and 1963.

Buyers could still add the Sports Roadster's tonneau by dealer order in 1964, the year the Thunderbird's fourth three-year installment arrived. In typical fashion, this body style was slightly revamped in 1965. But in 1966, the last of the fourth-generation 'Birds was treated to a few more styling updates in comparison to its Roundbird and Squarebird forerunners in 1963 and 1960, respectively. The final personal luxury T-Bird is easily identified thanks to noticeable trim adjustments at both ends. At a glance, these changes quickly set the 1966 Thunderbird apart from its 1965 and 1964 predecessors. This was also the last year for unitized body/frame construction, which had been a T-Bird trademark since 1958.

A fifth-generation Thunderbird, this one featuring body-on-frame construction, landed in 1967 with a thud. Longer and bigger all over, the 1967 T-Bird was also offered in four-door sedan form—with suicide-style rear doors. There was nothing personal about these luxury monsters. A restyle three years later added six more inches of length, this coming by way of a pronounced "beak" that easily identified the brief sixth-generation T-Bird run of 1970–71.

The seventh-generation Thunderbird grew truly huge in 1972 as Ford's luxury leader basically became a rebadged Lincoln Continental Mk IV. Four-door models were dropped this year, a season

The Thunderbird has certainly been many different cars for many different drivers over its 43-year career. Upsizing, downsizing, it didn't matter—they were all T-Birds. This 1979 'Bird was the last of its generation, a trimmed-down platform first offered in 1977. The Thunderbird was trimmed down even more in 1980.

marked by the production of the one-millionth Thunderbird.

Sportiness (and perhaps sanity) reappeared in 1977 when the slimmer, trimmer, eighth-generation Thunderbird debuted with yet another unitized body, this one shared with the midsized LTD II and Mercury Cougar. Sales skyrocketed to 318,140 that year, then an all-time high for the T-Bird line and a whopping 600 percent increase over the 1976 total. A new record of 352,751 was set in 1978.

And just when it looked like the Thunderbird was soaring like never before, a shift to the three-year-old Fox platform (Ford Fairmont/Mercury Zephyr) in 1980 brought the legacy back down to earth for its ninth-generation run. The downsized, boxy body did little to enhance the T-Bird image, nor did weak-kneed engineering. As if Ford's small-block V-8s of that era weren't bad enough, a six-cylinder was also introduced into the T-Bird mix for the first time. Status and prestige ebbed away during this weakly received three-year run.

A totally fresh image debuted in 1983. Sleek and rakish, the 10th-generation T-Bird was based on a revamped Fox-chassis platform. On top, this "Aero Luxury Car" cut through the wind to the tune of a 0.35 drag coefficient, not bad for a car that just one year before had featured all the aerodynamics of a brick. Performance was also enhanced by the midyear appearance of the popular Turbo Coupe with its forced-induction 2.3-liter four-cylinder engine. Buyers couldn't resist, and Thunderbird sales in 1983 jumped 170 percent compared to 1982's feeble results. "Aerobird" popularity remained strong throughout this generation's six-year life span.

An even more aerodynamic body draped over a new MN12 chassis featuring independent rear suspension drew

Forty years after one of the greatest American classics hit the streets, Ford rolled out its last Thunderbird for 1997. These bookend 'Birds are joined by a few of the high-flying machines of Kermit Weeks' Fantasy of Flight Aviation Museum in Polk City, Florida.

The Thunderbird's public farewell came in November 1997 at NASCAR's NAPA 500 in Atlanta. Here Mark Martin pilots his T-Bird for the last time. In 1998, he and the other Ford drivers traded their 'Birds for Taurus four-doors.

major raves in 1989 as the final member of the Thunderbird family bowed. And in place of the Turbo Coupe, which had disappeared along with the last Fox-chassis 'Bird in 1988, came the exciting Super Coupe powered by a 210-horsepower supercharged 3.8-liter V-6. Blown SC Thunderbirds were offered up only through 1995 while the MN12 platform rolled on in painfully similar form through nine model years. What made such grand headlines in 1989 became old news in a hurry as customers in the 1990s watched as basically the same old Thunderbird came and went year in, year out. The repetitive nature of the last-generation Thunderbird, however, shouldn't obscure the plain truth that all the MN12 'Birds were fine automobiles. To the end, they lived up to the Thunderbird name with pride.

Proof of just how well respected the 11th-generation T-Bird was came in 1989 courtesy of *Motor Trend* magazine, which honored the 1989 T-Bird SC as "Car of the Year." Two other Thunderbirds had previously driven home with "Car of the Year" trophies: the first Squarebird in 1958 and the upgraded 1987 Turbo Coupe. Another high-profile honor had come in 1961 when a Thunderbird convertible was chosen as the prestigious pace car for the Indianapolis 500.

Ford image-makers created special moments more than once to hype important dates in Thunderbird history. For example, all T-Birds in 1975 were considered 20th Anniversary models. Specially painted and trimmed models also appeared in 1980 and 1990 to mark the 'Bird's 25th and 35th birthdays, respectively. Dearborn didn't officially honor the 40th Thunderbird in 1995, but dealers were more than willing to stripe one up for you in full commemorative fashion. Whether or not Ford will let the car's 50th anniversary go by uncelebrated in 2005 is anyone's guess. Of course, they'll need to have another next-generation Thunderbird up and running first.

And you can bet when that happens horns will blow and confetti will fly.

Chapter 1

T-Bird for Two 1955-1957

Forty-three model runs–that's how many different T-Birds there have been. Few other long-running automotive nameplates have known longevity like that. Right now, only Chevrolet's Corvette can claim a longer production life.

Yet, despite those 43 successful years and 4.3 million cars sold up through 1997, many bystanders, innocent or otherwise, still only pay homage to Ford's first three Thunderbirds, the beloved two-seaters that sped off with so many hearts nearly a half century ago. "Littlebirds," "Earlybirds," or whatever you want to call them, the original Thunderbirds were, are, and always will remain the only T-Birds worth remembering as far as many fanatic followers are concerned. Indeed, in these minds, Ford stopped making Thunderbirds after 1957. Squarebirds,

Resemblances between the first two-seat Thunderbirds and the full-sized Ford line were made additionally apparent thanks to this trim treatment, which just missed regular production. Like Ford's standard passenger cars, the 1955 Thunderbird also would've probably been offered with two-tone paint. *Ford Motor Company*

Although a 1955 Thunderbird wearing serial number 100004 is now believed to exist, this black beauty, owned by longtime T-Bird enthusiast George Watts, is still recognized as the "first" Thunderbird. Its serial number is 100005. It has long been assumed that any cars built before Watts' were used for testing, not for public sale.

Roundbirds, Bigbirds, Aerobirds; it doesn't matter. The rest are impostors. All of them.

OK, so maybe there have been way too many fences built way too tall among the Thunderbird faithful over the years. Biases are still strong, even after 43 years. Why can't we all just get along? Perhaps because we really don't have to, at least not as far as honoring famous automobiles of the past is concerned. Thunderbirds have meant so many different things to so many different people during their historic run, that to group them under one banner would be unjust to all sides. Each individual faction deserves its own distinctive niche in automotive history. Yet, despite complaints that they're allowed to unfairly overshadow their successors, those first three T-Birds do receive the lion's share of nostalgic kudos, perhaps more so than all the following models combined.

Few automobiles, domestic or foreign, have ever inspired as much awe as the

Misguided assumptions that Ford's first Thunderbirds were sports cars weren't hindered in the least by the presence of a standard tachometer, located at the instrument panel's far left corner.

original Thunderbird. And even fewer have shown such long legs. More than four decades after the last of the two-seat breed became history, the first-generation T-Bird easily ranks right up at the top with any of the most recognized, respected, and revered cars to ever roll out of Detroit.

Even people who couldn't care less about cars will stop and smile when a two-seat 'Bird rolls by. That long-hood/short-deck profile, the portholes added in 1956, and those little fins that sprouted in 1957. You can't swing a dead cat by the tail and not hit someone able to identify these features at a glance. Not being able to do so would be un-American. Forget baseball, apple pie, and Chevrolet–red, white, and blue icons don't come much more patriotic than the two-seat Thunderbird. When Dearborn's movers and shakers first began building Thunderbirds, they may well have captured the very essence of what makes a great American car.

Great American car buyers in 1954 had never seen anything quite like it. More than one witness mentioned that the long-hood, short-deck look harkened back to Edsel Ford's classic Lincoln Continentals of 1940–interestingly, similar comparisons would be made in 1964 between the original T-Bird and Lee Iacocca's all-new Mustang. More specifically, an obvious connection could've been made between the first Thunderbird and its natural rival, Chevrolet's Corvette.

Some say the first T-Bird was a Corvette knock-off. Trying to find a truly original idea in Detroit has never been an easy task and remains especially tough today. Ford in the 1950s was no exception, nor was Chevrolet.

The whole two-seat phenomenon came into being thanks to a little overseas inspiration. American servicemen coming home from the war in Europe brought back more than medals. Some were transformed by their first experience seeing and driving a British sports car.

The affliction spread slowly at first. As the 1950s dawned, American drivers began to feel a new itch, an achin' for makin'

Thunderbird serial numbers began in the 100,000 range. And for years, this plate, the fifth issued, was the lowest known entry in that progression for 1955.

tracks, a need for speed, or at least the sensation of such. Top-down travel in a tight roadster that handled the curves with unprecedented confidence became the release, a new freedom. Imagining that a car could be fun and frivolous, and that going from point A to point B didn't have to be a chore, meant it could really be an adventure. But while the classic European sports car ideal *was* finding a home in this country during the early 1950s, not all that many Yankees were convinced it could survive here long.

Most drivers in this country were still turned off by the European roadster's inherently rude, crude nature. Sewing machine motors, leaky tops, pesky side curtains, uncomfortably close quarters, and few conveniences were all drawbacks in the American car market. And stingy sales figures showed that America really didn't want a sports car in the 1950s–a European-style sports car, that is.

Early Corvettes were rolling proof of just what the U.S. market *would* accept in those days. Chevrolet's intriguing two-seater nearly died after only two years on the road, due to its all-too-close resemblance to the bare-bones British sporting breed and its lackluster drivetrain, which consisted of a yeoman six-cylinder backed by a Powerglide automatic transmission. The Corvette didn't find real success until Chevy designers wised up and repackaged their fiberglass flight of fancy in a more user-friendly fashion. Also thrown in were more power, more performance, and more practicality. Life-saving additions included a V-8 in 1955 and a three-speed manual transmission very late that year. Exterior door handles, roll-up windows, and an optional removable hardtop joined the mix in 1956. Presto, "America's sports car" was born and Corvette drivers haven't looked back since.

Ford officials, along with legions of showgoers, had their first look at the original Corvette in January 1953 during the GM Motorama in New York. Who cared that the market for such a playboy's toy then accounted for about 1/4 of 1 percent of total car sales in this country? What

Unlike Corvettes, which initially came only with six-cylinder power, the Thunderbird was a V-8 cruiser from the get-go. Ford's Y-block V-8, introduced in 1954, fell right in place beneath the 1955 T-Bird's scooped hood. Displacement was 292 cubic inches.

mattered was the fact that Chevrolet had beaten Ford to the sports car punch. Even though the fledgling sports car field represented shaky economic ground, Dearborn decision-makers weren't about to idly stand by while the Corvette rolled out unchallenged.

One month after the Motorama show, word was sent down from the top at Ford to develop a comparable two-seater pronto. Ford General Manager Lewis Crusoe's team wasted even less time than Chevy's crew had while rushing the Corvette to market. Chief Engineer William Burnett oversaw the car's mechanicals, while William Boyer, under the direction of legendary designer Frank Hershey, did the bulk of the styling work. Together, they had a completed form all but finalized by February 1954. The big debut came at the Detroit Auto Show with a wooden mock-up that soon had both the press and the public clamoring for the real thing.

Motor Trend's Don MacDonald was among the first to herald the arrival of something truly new on the Detroit scene. He wrote that Ford appeared ready to not only meet Chevrolet's challenge, but to do it one better. Like

Contrasting this 1956 'Bird is a 1962 convertible. In 1956, output for the standard 292 V-8 was increased to 202 horsepower. *Steve Statham*

MacDonald, nearly all who saw the Thunderbird mock-up in Detroit recognized that the only aspect copied from the Corvette was the topless body with two-passenger seating and a short, 102-inch wheelbase. From the very beginning, Crusoe's plan had involved creating not a direct sports car competitor for the Corvette but an unquestionably fresh combination of youthful, sporty flair and outstanding prestige, all tied up in a package that was reasonably practical and wonderfully frivolous at the same time.

MacDonald's one-page April 1954 *MT* report, entitled "Thunderin' Thunderbird," explained that Crusoe had insisted that his designers stick with a metal body, not fiberglass, in order to keep manufacturing difficulties down while speeding production up. Clearly, Ford intended to build T-Birds faster, and in greater numbers than Chevrolet could mold up its Corvette.

The idea was not only to produce more cars, but to produce more car. "Perhaps the outstanding feature of the new Ford Thunderbird is the clever wedding of sports car functionalism with American standards of comfort," wrote MacDonald. Unlike the Corvette, the very first Thunderbird would have roll-up windows, a removable hardtop, and standard V-8 power. Chevrolet customers in search of fun behind the wheel of an automobile were still awaiting these warmly welcomed features.

Another Ford priority involved how quickly and easily the Thunderbird could be designed and built. Like the Corvette, which had many off-the-shelf passenger-car parts hidden beneath that innovative fiberglass shell, the first 'Bird relied on as

Noticeably new features for the 1956 Thunderbird included windwings for the windshield frame, cowl vents in the fenders, and the soon-to-be-popular "portholes" for the removable hardtop. Hardtops in 1956 came with or without portholes. They were essentially offered as a no-cost option.

FORD
FORD

A cleaner, revised grille/bumper layout announced the arrival of the third two-seat Thunderbird in 1957.

many "family ties" as possible to reach regular production. But, unlike the Chevy, Ford's two-seater also showed its passenger-line heritage on the outside. The fact that the original Thunderbird looked very much like a scaled-down 1955 Ford was no coincidence. Reportedly, the car's rapid-fire design process began with Bill Burnett's engineers simply cutting a standard 1955 sedan apart and welding it back together on that shortened 102-inch wheelbase.

From there, many styling cues followed closely along the lines of Ford's new full-sized look for 1955. From its hooded headlights to its round taillights, the Thunderbird was clearly every inch a Ford.

The easiest way to identify a 1956 Thunderbird is by its standard Continental kit spare tire carrier, which many felt didn't do the car any favors from a ride and handling perspective. The spare would go back inside an enlarged trunk in 1957.

Probably the most widely recognized classic T-Bird trademark is its "porthole" opera windows, first introduced as an option for the 'Bird's removable hardtop in 1956. They were so popular many believed they came standard. This is a 1957 hardtop.

Designers even considered using bodyside trim identical to the full-sized line–the so-called "Fairlane stripe." The thought was that two-tone paint options, similar to those used by the big Fairlanes, would then fit the little 'Birds to a "T" as well. At least one, and perhaps two prototypes were finished with the Fairlane trim, and an early advertisement in late 1954 even depicted the finalized Thunderbird form with this trim. Fortunately the idea was discarded at the last moment–the first 'Bird looked much better with less rather than more.

Early plans also hinted at using the Fairlane name–also the moniker of Henry Ford's estate–for the new two-seater. Instead, a corporation-wide contest was held offering $250 to the employee who could best name the breed. Stylist Alden "Gib" Gilberson took the prize, which ended up being a suit in place of the cash. A native of the American Southwest, Gilberson turned to familiar Native American mythology for his inspiration. Thunderbird was a legendary name for a soon-to-be-legendary car.

Production of Ford's legend began on September 9, 1954, with the Thunderbird officially going on sale to the public October 22. Teased for most of the previous year, customers couldn't get into dealerships fast enough to buy a car that left many wondering exactly what they were seeing. It looked like a sports car. It was small and light and fast, yet it impressed like a luxury car. It was also comfortable and had a much smoother ride than Chevy's Corvette.

Reporters weren't even sure what Ford had wrought. "The accusation that American car manufacturers couldn't build a sports car–even if they tried–is no longer valid," wrote Walt Woron of *Motor Trend.* "The first indication was the Chevrolet Corvette. And although the Ford Motor Co. is the first one to deny it, they have a sports car in the Thunderbird, and it's a good one."

Road & Track's editors, who preferred covering true sports cars over anything else, were not so quick to call a spade a spade. "The first truly American 'personal' car is the way the Ford Motor Company describes their new 2/3 seater, high-performance Thunderbird," began an October 1954 *R&T* report. "To the purist the Thunderbird has far too much luxury to qualify as a sports car, but even that group will find much of interest in the specifications of this car."

Most interesting was the power source, a bored-out version of Ford's 272-cid Y-block V-8. Featuring a single four-barrel carburetor, 8.1:1 compression, and dual exhausts, the standard 292-cid Thunderbird V-8 injected 193 healthy horses into the two-seat equation—when the base three-speed manual transmission was chosen. Adding the optional Ford-O-Matic automatic transmission also brought along a slight compression increase to 8.5:1, meaning output in turn

More optional underhood thunder was added in 1956 as the Y-block was enlarged to 312 cubic inches. With a single four-barrel carburetor shooting the juice in 1957, the 312 Thunderbird V-8 produced 245 horsepower.

A lengthened trunk allowed designers to hide the spare tire back beneath the rear decklid in 1957. Crowning the 1957 'Bird's new tail was a pair of humble fins atop each taillight. Powering this example is the rare supercharged 312 V-8.

went up 5 horses to 198. Performance was commonly listed as 0-60 in about nine seconds. The top end was right about 120 miles per hour.

Standard 1955 Thunderbird equipment also included a removable fiberglass hardtop, at least it did after some early confusion was cleared up. Apparently some early cars came standard with no top at all, since both the lift-off hardtop and folding soft top were initially listed in factory paperwork as options along with conveniences like power windows and a power seat. Eventually, however, the hardtop was included in the base price of about $3,000. Adding a convertible top in place of the hardtop cost an additional $75. Price for both tops was $290.

With the top off or down, a T-Bird driver received a decent dose of sports car feel, an impression backed up by all the horsepower supplied beneath that polite little hood scoop. With the hardtop on, passengers were comforted in the knowledge that they didn't have to arrive at the club with the Don King look. Your hair not only stayed in place, it also remained dry. Such sporty performance had never come so classy and convenient before. Maybe that's why the first Thunderbird outsold Chevrolet's 1955 Corvette by a 23-to-1 margin.

Plans to sell 20,000 T-Birds in 1956 proved a bit optimistic. Nonetheless, 15,631 'Birds out the door the second time around was certainly nothing to sneeze at, and that for a car only slightly updated. Noticeably new features for the 1956 Thunderbird included windwings for the windshield frame, cowl vents in the fenders, and the soon-to-be-popular "portholes" for the removable hardtop. Hardtops in 1956 came with or without portholes as they were essentially offered as a no-cost option.

To make room in the trunk, designers also moved the spare tire outside into a classic-style Continental carrier mounted on the rear bumper. Other more subtle changes in back included moving the dual exhaust exits to the bumper's corners (since the Continental kit now blocked the position used in 1955) and slightly revising the taillights.

A floor shifter and engine-turned dash face enhanced the Thunderbird's sporty image from behind the wheel. The small circular device suspended from the dashboard below the radio is the memory control for the optional Dial-A-Matic power seat. A driver could set this control for height (five choices) or front-to-back preferences (seven choices). When the key was turned off, the seat moved rearward to allow easy exit and re-entry. Once the key was turned back on, the seat would then return back to its preset position. This option was offered only in 1957.

Ford briefly turned up the heat for the Thunderbird early in 1957, offering two race-ready engine options. The first choice added twin four-barrel carburetors atop the 312 Y-block pushing output to 270 horses.

Underhood upgrades for 1956 were both subtle and serious. First, output for the 292 V-8 was increased to 202 horsepower. The really big news, however, was the introduction of the enlarged 312-cid Y-block, which was available with either 215 or 225 horses. Chevy's Corvette that year was also advertised with 225 horsepower. By the end of 1956, there were nearly four times as many Thunderbirds on the road as Corvettes. Consider also that Chevy had a two-year head start on production.

In 1957, the third-edition Thunderbird's spare tire was put back inside the car as the trunk was lengthened some six inches. Helping announce this restyle were outward-canted fins atop each rear quarter. These were courtesy of Frank Hershey, who had played a major role in GM's introduction of the fin fad while working at Cadillac in 1948. Up front, a new and larger grille and a revised bumper came along with the last of the two-seat T-Birds. Throw in a few other minor updates and you have what most consider to be the best of the breed. Buyers apparently agreed. Ford did finally break the 20,000-unit sales barrier in 1957.

Adding to the 1957 Thunderbird's attraction was even more strength beneath that long, long hood. Output for the 292 and 312 Y-blocks that year increased to 212 and 245 horsepower, respectively. And that wasn't all.

From the beginning, Thunderbirds' V-8s had used single carburetors to handle fuel metering chores. Then, late in 1956, Ford tried a dual–four-barrel intake for the 312 Y-block, this addition reportedly boosting horsepower to 265. In

1957, the twin Holley carb option was offered again as part of the "E-code" Thunderbird Special V-8 package, rated at 270 horses. Adding the "NASCAR kit" cam upped the E-Bird's output to 285 horsepower. But if that didn't light your fire, there was more.

The hottest Thunderbird V-8 in 1957 used only one Holley four-barrel. Attached to that carburetor, however, was a McCulloch centrifugal supercharger supplying as much as six pounds of boost by way of its belt-driven impeller. Advertised output for the "F-code" supercharged 312 was 300 horsepower; 340 with the solid-lifter NASCAR cam. Only 208 F-Birds were built, 14 of those being the "D/F" versions built specially in January 1957 with NASCAR tracks in mind. Estimates put twin-carb E-Bird production at about 500.

Both models E and F came and went quickly during the first half of the 1957 model year as Ford was forced to change its plan for taking its high-flying Y-blocks racing. In June 1957, the Automobile Manufacturers Association issued what amounted to be a ban on such "unsafe" shenanigans. Initially offered for both the Thunderbird and Ford's passenger-car line, the dual-carb and supercharged 312 V-8s were immediately canceled right about the time of the AMA decree, victims of a newfound need in Detroit to curb what some considered to be a horsepower race running out of bounds.

Ford's two-seat T-Bird followed those high-powered engines into history a few months later, much to the dismay of so many Thunderbird lovers who never were able to warm up to the four-place 'Birds that followed.

Top power in 1957 came from this supercharged 312, which used a belt-driven blower to boost output to 300 horses. Adding an optional solid-lifter cam reportedly unleashed another 40 horsepower.

Thunderbird
COAST
GUARD

Chapter 2

Four is More 1958-1966

Time sometimes has a strange way of wounding all heals. Consider the fate of the Squarebird, a car totally undeserving of the slings and arrows hurled its way by modern-day second-guessers. Sure, a great number of the people who had fallen madly for the two-seat Thunderbird were wholly disappointed when room for two more was added in 1958. Yet, those complaints aside, the first four-place T-Bird still emerged as an unqualified success, perhaps even an overnight sensation.

A refocused target market thought so. After a late introduction, the 1958 Thunderbird managed to establish a new sales standard for the breed–and this in a recession-wracked year of disaster when no one in Detroit was selling cars. No one but Ford, that is. The redesigned four-seat 'Bird was one of only two American models in 1958 to lay claim to a sales increase, in this case a whopping 76 percent jump

It may have broken purists' hearts, but the first "Squarebird" saved the bloodline from extinction in 1958. The totally redesigned 1958 Thunderbird used unit-body construction in place of the body-on-frame platform found beneath the two-seat 'Bird.

Designers saved the Thunderbird from an early grave by widening its appeal in 1958. Key to this new direction was adding seating for two more passengers. A nonstock radio has been added by this 1958 T-Bird's owner.

over 1957's results. Depressed economy be damned, Dearborn successfully rolled out 37,892 Thunderbirds in 1958. Another 67,456 followed in 1959, then 92,843 in 1960–the latter a record high that would stand until 1977. Clearly something was right about the car.

Most of the motor press in 1958 supported this claim to a great degree. According to *Motor Life*, the 1958 T-Bird was "probably the most modern American car in production." In the words of *Motor Trend*'s editors, it was "a car that combines safety with performance and comfort with compactness." The *MT* staff was so impressed, they picked the first Squarebird as their "Car of the Year" for 1958. A second "Car of the Year" trophy wouldn't come Thunderbird's way for another 30 years.

All that pomp and circumstance, all that sales success, yet the four-place Thunderbird still commonly gets a bum revisionist rap. Purists never have forgotten or forgiven. Inspired by their often loud cries of "foul," a car-loving public concerned more with nostalgic memories than historic fact has bought into the belief that Dearborn's decision to repackage the T-Bird in 1958 was a mistake. Stop any motorist with fuzzy dice hanging from the rearview mirror of his car today and ask him what he thinks. Chances are he will regurgitate a familiar line, "Ford stopped building Thunderbirds in 1957."

In truth, Dearborn probably would have done just that had Robert McNamara not come to the rescue with a bigger,

A three-year production run remained a Thunderbird practice into the Squarebird years. After 1960, the first of the four-place T-Bird bodystyles was replaced by the radically updated "Roundbird," which in turn rolled on through three more years. This convertible Squarebird is one of 11,860 built for 1960.

Standard Squarebird power came from Ford's new 352-cid FE-series V-8. As it had in 1958, this 352 Interceptor V-8 was still producing 300 horsepower for the 1960 Thunderbird.

better idea for 1958. Later criticized for his conservative ways, Ford's sales-conscious division chief was the main man behind the Thunderbird's radical redesign. "He fought with the board of directors to get it, and he won the fight," wrote designer Bill Boyer later in reference to the four-place 'Bird proposal.

McNamara did have eyes. He could see the limited appeal of the two-seat Thunderbird, and he could also read a balance sheet, where the ink wasn't anywhere near as black as Ford officials had hoped, despite three years of increasing sales. Many Ford officials were already doubting the T-Bird's ability to sustain extended flight as early as December 1954. Once McNamara rose to the general manager's seat not long afterward, he almost immediately took up the cause of keeping the Thunderbird alive at all costs. It was clear to him that a 'Bird of another feather would be required to carry on the breed, and soon.

Widening the car's appeal was the obvious answer, and increasing seating was the key to widening that appeal. Two may have been cool, but four was more. More room for more people to fit inside. More reasons for more people to rush into Ford dealerships. More sales. More Thunderbirds.

Most at Ford recognized a bigger 'Bird was not that far down the road even as the first Littlebird was turning heads everywhere in 1955. Product planner Tom Case even proposed a two-model lineup where the two-seater would continue in production, joined by a four-seat running mate. McNamara, however, shot this idea down. It would be four-place or no place for the future T-Bird. A full roof would also enclose the car for the first time as part of the plan to bring the Thunderbird closer to the mainstream.

But what about the Thunderbird's sports car image? Or its "personal luxury" place in Ford Motor Company's pecking order? Wouldn't more seats, more room, and more size negate these impressions? *Road & Track* came forth with an immediate answer to these questions in its February 1958 issue:

"While sports car fiends generally have not been too kind to the Thunderbird (an understatement), they should remember three things: (1) The Ford Motor Company has never called the T-Bird a sports car; it is a personal car, if you

In July 1960, Ford, working in concert with the Budd Body Company and Allegheny-Ludlum Steel, produced two experimental Thunderbirds on the Wixom assembly line using complete stainless-steel structures. They were the last two Squarebirds built, as officials knew the "less cooperative" stainless steel would ruin the body stamping dies. This stainless 1960 T-Bird still survives at Allegheny-Ludlum's Pennsylvania works. *Gene Makrancy*

The first Roundbird was chosen as the prestigious pace car for the 1961 Indianapolis 500, its golden finish helping mark the 50th anniversary of the fabled Brickyard's first Memorial Day race. *Tom Glatch*

please. (2) Most Americans like luxury and will never accept the starkness of a true sports car. (3) Many T-Bird owners are willing and frequent converts to the fun-driving features of genuine dual-purpose sports cars."

There it was. The Thunderbird never really was a sports car, it was a sporty luxury car. More to the point, it was a sporty personal luxury car. And since Ford people basically defined "personal luxury," they could build it in any fashion they wanted. "As a result of their own surveys," explained *Road & Track*'s report, "Ford decided that their personal car should be a four-seater."

Ford basically considered the four-place Thunderbird hardtop to be a refinement of the personal luxury idea, even though the car was considerably less personal than its two-seat forerunner thanks to its bigger body. Most early responses to the 1958 T-Bird included references to the car's clear family ties to Lincoln's truly huge Continentals, a connection that ran deeper than casual witnesses realized since both models relied on Ford's new unitized body/frame construction, and thus shared the same production line at the equally new assembly plant in Wixom, Michigan.

On the one hand, exterior impressions of the 1958 T-Bird were quite "Lincolnesque." They clearly represented a distinct departure from the breed's sporty beginnings. As *Motor Life*'s reviewers explained it, "the ['58] Thunderbird boasts a new maturity which past enthusiasts may call conservative."

Looks, on the other hand, were deceiving. Despite the fact that the 1958 T-Bird appeared to dwarf its predecessors, it still retained some of that warm and fuzzy feeling first offered in 1955. Compared to Detroit's incredibly huge, overstuffed luxury showboats of the late 1950s, the 1958 Thunderbird was very much a "personal car." Its four-place bucket seat interior with console was completely comfortable and warmly welcoming to drivers looking for a little pizzazz and prestige in a relatively polite package. It was also seriously cool, with a sporty feel behind the wheel that remained a T-Bird trademark, even though the car now wore a full time steel roof and looked more like a luxury liner on the outside.

Making room for two more riders required a stretched 113-inch wheelbase for the Squarebird's unitized platform—11 inches longer than the Littlebird's chassis. Overall length went up two feet, width increased by six inches and weight jumped up about a half a ton. An exceptionally low 5.8-inch ground clearance kept roof height down, something designers planned all along. Trends of the day dictated longer, lower, wider lines for the 1958 Thunderbird, which in turn accentuated the car's newfound mass.

Two new variations on the Thunderbird theme were introduced in 1962: the prestigious Landau hardtop and sexy Sports Roadster. This black Landau is one of 12,139 built for 1963.

Along with a vinyl roof and classic landau bars, the Thunderbird Landau also featured various interior dress-ups. Simulated walnut trim was part of the deal.

Also new for the 1958 platform were coil springs in back. These were added in the hopes of incorporating the division's "Ford Aire" air suspension. However, once those air springs fizzled beneath Ford's standard line in the fall of 1958, they were nixed from the Thunderbird layout as well.

Dealer introductions of the four-place Thunderbird began on February 13, 1958, more than four months after other "new-for-'58" Fords. A total redesign typically took longer than planned. Official production at Wixom had begun on January 13 and initially only included hardtop models. With approval for a topless Squarebird not coming until May 1957, convertible production was also delayed. The first 1958 T-Bird convertible–with semiautomatic power top operation–didn't appear on showroom floors until June 1958.

Power for the Squarebird came from Ford's new FE-series V-8. Displacement was 352 cubic inches; output was 300 horsepower. Engineers reportedly considered offering the 361-cid Edsel V-8 as an

In February 1963, Ford introduced a special limited-edition Landau model in Monaco, where Princess Grace was given the first such example. Only 2,000 were built, all painted white with white leather interiors and deep maroon vinyl tops. *John Lee*

MONACO

Stunning Kelsey-Hayes wire wheels were included as part of the Sports Roadster package along with that distinctive tonneau cover that converted the car back into a classic two-seater. Only 455 Sports Roadsters were built for 1963, including this red-hot example.

option, and some claim the 375 horsepower 430-cid Lincoln V-8 did make the Thunderbird list late in the year. While the former never got beyond the planning stage, the latter did at least appear in prototype form for a *Motor Trend* road test in 1958. However, actual production of 430-powered 'Birds that year is doubted.

The big 430 V-8, in this case rated at 350 horsepower, did become an official Thunderbird option in 1959, when a second Squarebird appeared in nearly identical fashion, save for a new grille and different trim for the bodysides and taillights. After spending $5 million rebuilding the T-Bird for 1958, Ford was only able to set aside another $2 million to update the second-generation Thunderbird through its planned three-year run. Unfortunately, design work on the "afterthought" 1958 convertible had taken a major bite out of that future budget, thus the reason for so few upgrades.

More minor trim adjustments and a switch from four taillights to six announced the 1960 Thunderbird's arrival. Other improvements during the Squarebird's run included the addition of a fully automatic convertible top late in 1959 and the introduction of this country's first postwar sunroof, a manually controlled hardtop option offered in 1960. Additionally, the coil spring suspension used in 1958 was traded for traditional leaf springs for 1959 and 1960.

Three-year generations continued to be a Thunderbird tradition through the four-place run. The last Squarebird was

Thunderbird interiors never lacked appeal, and the 1963 Sports Roadster was no exception. Notice the console-mounted tachometer.

One of the hottest Thunderbird engine options of all time was the M-code 390-cid V-8 fed by three Holley two-barrel carburetors. Introduced in 1962, this triple-carb FE-series big-block put out 340 horsepower. This M-code application is beneath a 1963 Sports Roadster's hood. Only 37 340-horse Sports Roadsters were built that year.

hustled out the door early in July 1960 to make room for the restyled, aeronautically inspired Roundbird, which rolled on, once more, in nearly identical fashion up through 1963. The last three-year group of the four-place family featured even more modern, truly crisp styling, which in turn led to a new rise in popularity. The 1964 Thunderbird outsold its 1963 predecessor by almost 30,000 units, nearly setting a new sales standard in the process. Final production that year was 92,465. At 69,176, Thunderbird sales were still healthy two years later for the four-place 'Bird's final flight, but 1966 was also the last time a convertible T-Bird was offered.

Technical highlights from the third- and fourth-generation runs included a typical displacement escalation, first to 390 cubic inches in 1961, then to 428 cubes for another FE-series big-block V-8 introduced as an option in 1966. An optional "Swing-Away" steering wheel was introduced in 1961, an intriguing feature that enhanced a wonderfully sporty interior that was all but unmatched in Detroit for its sexy style. Front disc brakes and sequential taillights also became standard equipment in 1965.

Milestones included an appearance at "The Brickyard" in 1961 as the prestigious pace car for the Indianapolis 500. Two new models were unveiled in 1962, one created to enhance the Thunderbird's luxury reputation, the other intended to revive memories of the car's two-seat heritage.

The first, a Landau hardtop, added a standard vinyl roof with classically inspired "landau irons" on each rear pillar. Simulated walnut interior appointments became part of the standard Landau hardtop deal in 1963. A special-edition "Princess Grace" (or "Monaco") Landau was also introduced in February 1963. All 2,000 built featured white paint, Rose Beige vinyl roofs, white leather upholstery, and simulated rosewood interior accents. A special serial number plate on the dash was the crowning touch. A second limited-edition Landau hardtop model appeared in March 1965, this one featuring Emberglo paint with matching wheelcovers and interior appointments. A dash-mounted serial number plate again marked these cars, of which 4,500 were built.

The second Thunderbird variation that debuted in 1962 came in response to all those complaints first heard in 1958. Amazingly, two-seat purists' protests didn't all fall on deaf ears. As early as 1960, work had begun on a new type of sporty T-Bird, thanks to Ford Marketing Manager Lee Iacocca, who apparently felt that those who helped make the first Thunderbirds a soaring success shouldn't be left flat. More to the truth, Iacocca was bombarded with requests from dealers for a return to the two-seat theme. The result was the Sports Roadster, the sexy topless T-Bird with that tight-fitting tonneau that converted a four-place Roundbird into a two-seat tourer.

The idea wasn't exactly new. As early as 1958, New Jersey Ford dealer Bill Booth had been manufacturing fiberglass rear seat tonneau covers for Squarebird convertibles. With the groundwork established, it didn't take long for Ford designers to try the same trick on their own. Credited primarily to stylist Bud Kaufman, the Sports Roadsters' fiberglass tonneau fit almost like a glove and at the same time allowed the convertible top to operate unheeded. Twin headrests, which were more stylish than functional, were incorporated into the cover. The package

While the Sports Roadster model was dropped after 1963, Thunderbird buyers could still dealer-order a similar tonneau cover in 1964. *Ford Motor Company*

RK
U.S.AIR FORCE
54785

also featured exclusive fender emblems, a dash-mounted grab bar, and four dazzling Kelsey-Hayes wire wheels.

Sports Roadsters were offered in both 1962 and 1963, each wearing heavy price tags: $5,439 in 1962, and $5,563 in 1963. Those numbers help explain why buyers stayed away in droves. Production was only 1,427 in 1962, a mere 455 the following year.

Even more rare were the "M-code" Sports Roadsters. In January 1962, Ford officially introduced the M-code engine option, a $242 package that exchanged the standard Z-code 390 V-8's four-barrel carburetor for three Holley two-barrels. Output for the "390-6V" big-block was 340 horsepower, increased by 40 horses from the Z engine. And along with various engineering tweaks, the M-code Thunderbird V-8 also wore a nice dose of chrome-plated dress-up pieces.

M-code T-Bird production was meager to say the least–145 in 1962, 55 in 1963. Sports Roadsters made up the bulk of these installations (the option was available for four-place 'Birds, too), totaling 120 in 1962, and 37 in 1963. Even though the six-barrel Sports Roadster probably ranks as the sportiest Thunderbird ever, its limited-edition presence allowed it to escape the attentions of most casual observers. Too bad–this was the one member of the family that truly did put the thunder in Thunderbird.

Ford's final four-place Thunderbird body debuted in 1964 to the apparent delight of customers who couldn't get enough. Sales soared from 63,313 for the last Roundbird to 92,465 for the restyled 1964 'Bird.

T-Birds in Pop Culture

Bestowing larger-than-life status on favored subjects is a practice Americans can't resist. Call it a need to look up to something, or classify it as an addiction to nostalgia. Whatever the reason, we tend to hand out immortality I.D. cards to people and things as if we have the power to restore life to the former, or create it for the latter. In both cases, that artificial life called fame almost always grows in prominence as time marches on.

Marilyn Monroe wasn't just an actress (some feel she wasn't even an actress), she was the personification of sex—soft, slow, slithering, steamy sex. In life, Norma Jean Baker was the "It Girl" for the 1950s generation. In death, she has become so highly revered by the baby boomer crowd, men and women alike, that her image has been purified almost to the point of restoring her virginity. No one today dares imagine her photographic phantom in unclad fashion, something *Playboy* readers had actually witnessed in 1953.

As in Marilyn's case, the jury is also still out on Elvis Presley. Sure his songs were fun, his attraction was electric, but his lyrical contributions to musical history were far from classic. His celluloid efforts were even less so. Yet his ghost continues to stalk us at convenience stores all across the country. Elvis had already reached a larger-than-life place (in more ways than one) during his last few years. After his alleged death, he really began throwing his weight around. To most fans today Elvis wasn't just a rock star, he was (and still is) The King. "Larger" isn't a big enough word.

Then there's the classic Thunderbird. Nostalgic icons may not come much larger, certainly not in the automotive realm. This wasn't just a car, it was an E-ticket ride to the fountain of youth. A two-seat T-Bird worked better than a toupee as a youthful tonic in 1955. Nearly a half century later the car still revives fond remembrances of yesterday. Even fans of the fabled '57 Chevy have to admit that their fave takes a backseat to the "Earlybird" in the race down memory lane. Pick up a "Lost in the '50s" T-shirt today and chances are it will immortalize one of at least three major icons — Norma Jean, The King, or a '57 T-Bird.

Ford's first Thunderbirds probably didn't need much help becoming hallowed pieces of Americana automotive art. Their unique market niche, their playful persona—all this and more instantly guaranteed that we would never forget these cute little cars. Like Packards, Duesenbergs, and Tuckers before them, some '55–57 Thunderbirds may have temporarily suffered the ignominious fate of sitting soaped-up on Honest John's used car lot in the early 1960s. But it wasn't long before they were rescued and restored to their proper place in automotive Valhalla.

Elvira, B-movie queen, is the proud owner of a ghoulishly customized 1958 Thunderbird convertible. Her Macabre Mobile, created by George Barris, features a spider web grill, skull and bones hubcaps, and a chain steering wheel with pentagram spokes. It played a prominently role in her 1980s film, *Elvira, Mistress of the Dark. Neil Preston, courtesy Queen B Productions*

These cars were stars even before they first began joining up with pop culture forces in the 1960s, meaning a quick trip to immortality was made even easier. Once preserved forever in song and on the silver screen, the Thunderbird legend achieved a wider scope—a mass appeal for more casual observers who otherwise may have never considered it possible for a car to become a pop icon.

Soon to be icons themselves, the Beach Boys first turned to Dearborn's two-seater for lyrical inspiration in 1963. When songwriter Brian Wilson needed a measuring stick to demonstrate the rubber-burning prowess of a '32 Ford hot rod he chose a T-Bird. "Just a Little Deuce Coupe with a flathead mill," crooned the Wilson brothers and friends. "But she'll walk a Thunderbird like it's standing still." Who cared that a T-Bird really wasn't that hot of a car? Girl- and boy-crazy Beach Boys fans never noticed—they truly didn't "know what I got"—nor did they care. Most of them never set foot on a surfboard, either.

What the fans did know was how to have loads of fun, at least until "daddy took the T-Bird away." The Beach Boy's 1964 hit single, "*Fun, Fun, Fun*," better reflected the Thunderbird's place in the cool world. A performance reputation was purely the product of mirrors. Status, prestige, and pizzazz, on the other hand, were all proven quantities. In this instance Wilson hit the nail right on the head. A two-seat T-Bird was the perfect rich man's toy, whether the playmate in question was a daughter, a wife, or a mistress.

This classy chassis image was perfect for both movies and television. A red two-seat T-Bird fit Robert Urich like a glove for his role as detective Dan Tanna in ABC's "Vegas," aired from 1978 to 1981. Sexy assistants, life-in-the-fast-lane Las Vegas nights, and a too-cool-for-school Thunderbird—what else could a man want?

In Curt Henderson's case, the answer to that question came in the form of a '56 Thunderbird. Easily the most renowned starring role for a T-Bird came in George Lucas' 1973 hit movie *American Graffiti*, when Henderson, played by Richard Dreyfuss, stumbles upon the perceived love of his life, Suzanne Somers, while cruising the strip the night before he is to fly off east to college. "The most perfect, dazzling creature I've ever seen" continues to taunt Henderson from her virgin white T-Bird right up to the movie's end. Whether she was the young trophy wife of a wealthy jeweler or a high-priced call girl mattered not at all. "I just saw a vision, I saw a goddess," was Curt's claim. But would he have been equally dazzled had Somers been riding in the backseat of a '58 Edsel four-door like him?

Four-place Thunderbirds, to a lesser degree, also became cogs in Hollywood's image machine. On television, Paul Drake, Perry Mason's soft-spoken, broad-shouldered private investigator, drove a Squarebird convertible with style and flair on the rare occasions when he wasn't standing around Mason's office taking up space. Another Squarebird later accompanied Elvira, Mistress of the Dark, in the 1980s, if only because it could be customized so easily to reflect its own dark side. Who was looking at the car though?

Poor little rich boy Robert Conrad showed little respect for his Roundbird convertible while playing bump and run with Troy Donahue in the 1963 film *Palm Springs Weekend.* But Conrad's destructive driving paled in comparison to what Geena Davis and Susan Sarandon did to a '66 T-Bird droptop in the 1991 movie *Thelma and Louise.* Just because the name was Thunderbird didn't mean the car could fly.

Fortunately, the T-Bird legacy hasn't suffered the same fate. Like Elvis, it will never die, not as long as art continues to perpetuate life.

Chapter 3

A Really Big Show 1967-1979

Without a doubt, the Thunderbird tale ranks as one of the most diverse, if not disjointed, in American automotive history. To say the T-Bird lineage has featured many different cars for many different drivers is an understatement. One-time rival Corvette has managed to stay on the road for as long as it has by doing one thing really well. It was, is, and always will be, "America's sports car." Thunderbird, on the other hand, has owed its lengthy career to Ford's frequent willingness to make alterations. Surviving for 43 model runs in a marketplace full of fickle car buyers was a matter of adaptability. Change, sometimes radical, sometimes subtle, was a T-Bird trademark from the beginning.

Purists can stomp their feet all they want in reference to the original two-seater's all-too-short career. If Dearborn designers had not widened the T-Bird's

'Bird watchers who thought Ford had gone one step beyond in 1958 were jolted again in 1967. Adding two more places was one thing. Two more doors really changed all the rules.

appeal by doubling its seating capacity, the legacy undoubtedly would have ended right then and there–three years and out. Unlike Chevrolet, Ford simply wasn't willing to continue pouring its best efforts into a project that could only offer a return of 15,000 to 20,000 unit sales a year. Market realities ruled the auto industry. Keeping the Thunderbird flying beyond 1957 meant more was in order. More car. More riders. More sales.

But if Thunderbird followers were shocked when Ford added two more seats in 1958, they had another surprise coming in 1967. Luxury was still the Thunderbird's main attraction, but it was no longer so personal. Not even close. Times had clearly changed and the totally new fifth-generation Thunderbird was rolling proof . . . big rolling proof.

Actually, the 1967 Thunderbird only *appeared* larger than life. Its imposing image overshadowed specification realities. Styling chiefs David Ash and Bill Boyer saw to that by sculpting a thoroughly modern, angular body featuring lines that looked longer, lower, and wider than they really were. At 115 inches, the wheelbase was only two clicks longer than the 1966 model. The overall length also compared closely, 205.4 inches for 1966 against 206.9 inches for 1967. The overall width was the same. And as bulky as the 1967 'Bird appeared to the eye, in truth it weighed, almost amazingly, less than its

What a difference a year makes. In 1966, Ford built the last Thunderbird convertible. In 1967, the four-door Landau hardtop debuted as sexy sportiness was exchanged for status-conscious plushness. Dig those suicide doors in back.

forerunners–amazingly because at a glance the fifth-generation Thunderbird looked like twice the car in comparison to the 1966.

This apparent violation of physical laws can be explained by what lay beneath that seemingly expansive skin. The unitized body construction used from 1958 to 1966 was traded for a traditional full frame in 1967. Sure, that frame was reinforced with extra weight-gaining members to guarantee a quiet, sure ride. And the 1967 bodyshell itself was also beefed to help meet those ends. Yet the entire battleship-steady package–as rigid and roomy as it was–still weighed less than what came before, because unitized construction required a significant amount of body bracing to make up for the lack of a true frame. Cost cutting also played a role in the decision to switch back to body-on-frame construction–a unitized body was more expensive to manufacture.

The 1967 Thunderbird's excellent ride was also enhanced by a return to coil-spring suspension for the rear wheels, something earlier T-Birds had briefly used in 1958 only. Two-seat 'Birds and those built between 1959 and 1966 had all used parallel leaf springs in back.

New-for-1967 features, along with that palatial body, included trendy hideaway headlights. Up front, the 1967 Thunderbird was all grille. In back, it was all taillight. And down the bodysides it looked like it was all wheel, even though rim diameter, at 15 inches, remained a carryover from 1966. Full, semicircular rear wheel openings contributed to this effect due to both their definitely enlarged size and the fact they weren't enclosed by fender skirts, as had been a T-Bird practice from the outset in 1955.

The really big news for 1967, however, involved the Thunderbird's revised model lineup. In place of the deep-sixed convertible bodystyle came a new four-door sedan, a definite departure from the previously popularized personal luxury theme. Ford officials were especially proud of this break from the existing T-Bird ideal. Promotional paperwork played up the car's newfound place in Ford Motor Company's prestigious pecking order. "Heretofore, a buyer of Ford products had no place to go between the Mercury Park Lane and the Lincoln Continental," explained Ford's 1967 sales book. "Now he has Thunderbird."

Two extra doors–mounted in rearward-hinged "suicide" fashion–were an appropriate addition for this fresh brand of T-Bird status, at least in Ford men's minds. Others were not so easily impressed. "The idea of adding two more doors to the T-Bird is being treated like the invention of the cotton gin by Ford, but the change is hardly worth the hoopla," claimed a *Car and Driver* review.

Customers apparently thought differently. Production of the 1967 four-door Landau sedan (which rolled on a lengthened 117-inch wheelbase) reached 24,967. Compare that figure to the 1966 Thunderbird convertible's 5,049 production total and you begin to get the picture. Sales success talked, diminishing images

Trendy hideaway headlights became a Thunderbird feature for the first time in 1967.

walked. Clearly, trading sexy sportiness for prestigious practicality proved to be the wise move, slings and arrows from early 'Bird devotees notwithstanding.

The Thunderbird's fifth generation, per Ford tradition, lasted three years. Production was 77,956 in 1967, then the third highest total in T-Bird history. Additional highlights of the 1967–69 run included the addition of a standard front bench seat in 1968 that upped seating room to six. A 1969 headline-making addition was the introduction of a new standard Thunderbird engine, the 429-cid Thunder Jet V-8. In 1967 and 1968, a 390 FE-series big-block had been standard, with a 428 FE available as an option. In 1969, the only engine available was the new, cleaner-running 385-series 429, rated at 360 horsepower.

Considerably less well known, yet deserving of historical note, were the five "Apollo Thunderbirds" built by Dearborn Steel Tubing in 1967. Home to Ford's Thunderbolt Fairlane factory drag cars of 1964, Dearborn Steel Tubing dressed up these five two-door T-Birds as part of a promotional effort for the renowned Abercrombie & Fitch sporting goods firm. All five were painted metallic Apollo blue with a matching blue vinyl top. A complete list of options was complemented by an electric sun roof and quartz iodine driving lights. Unique touches included blue-chromed wheelcovers, golden Thunderbird grille badges, chromed door jambs, special side marker lights, and exclusive gold-tinted landau bars. Interior luxury was enhanced with, among other things, a radio, telephone, and a Philco television.

Priced at about $15,000, the Apollo Thunderbirds were displayed at Abercrombie & Fitch stores in Miami, Palm Beach, New York, and Chicago. The fifth model, intended for San Francisco, was destroyed on the way west. According to the International Thunderbird Club, at least two of the Apollo Thunderbirds have survived and are in collectors' hands.

Back in the regular-production world, a sixth-generation Thunderbird debuted in 1970, the product of the presence of short-term Ford President Semon "Bunkie" Knudsen, a General Motors defector who had brought much GM-style thinking to Dearborn when he had jumped across the fence in February 1968. Way too much of a mover and shaker for Henry Ford II's taste, Bunkie was quickly fired in September 1969, but not before he had been able to make a few changes. One of these involved adding a Pontiac-style "beak" to the Thunderbird that alone lengthened the car by six inches.

Like Bunkie Knudsen, the sixth-generation T-Bird was not well received and only lasted two years. After 50,364 were sold in 1970, production dropped like a rock to 36,055 in 1971, a signal that yet another change was in order.

Although some critics applauded the 1967 T-Bird's interior for the way it toned down the "airplane-pilot syndrome" common to earlier models, the driver's position still resembled an aircraft cockpit, thanks mostly to its large, wraparound console. In 1967, the Swing-Away steering wheel became a Tilt-Away as it moved to the side and tilted up to allow easy access.

The one-millionth Thunderbird was built at Ford's Pico Rivera plant in Los Angeles on June 22, 1972. After being used for a year by the Classic Thunderbird Club International's best of show winner, it was purchased by collector George Watts. Bob Peterson bought the car from Watts in 1985. *Bob Peterson*

With Lee Iacocca now in the top seat at Ford (since 1970), the Thunderbird was treated to a complete makeover for 1972 to undo what Bunkie had done. Iacocca's 'Bird would end up being the largest yet, thanks to the fact that it shared the Lincoln Continental Mk IV's platform. The seventh-generation Thunderbird, a luxury showboat to say the least, rolled out on a truly long 120.4-inch wheelbase. The overall length was an aircraft-carrier-like 216 inches, while the total width was 79.3 inches.

With weight also increased, Ford engineers in 1972 introduced a new optional V-8 for the Thunderbird, a 460-cubic-inch monster that shoved aside the Squarebird's old 430 to become the largest T-Bird engine ever offered. Net output rating for the low-compression 460 big-block was 224 horsepower, compared to 212 horses for the standard 429 V-8.

The year also marked a return to a singular model offering since the four-door sedan was dropped. Roomier and more luxurious than ever before, the 1972 two-door Thunderbird brought buyers back into the fold in impressive fashion. Sales that year soared by nearly 70 percent over 1971's results to 57,814. Another 87,269 Lincoln-'Birds were sold in 1973, a new third-place sales standard for the Thunderbird heritage.

Ford had sold its one-millionth Thunderbird the year before, an event that didn't go by unnoticed. A special photo opportunity, complete with an obligatory banner, was arranged as the car rolled off the Los Angeles assembly line on June 22, 1972. Special touches to the car itself included custom gold paint set off by a white vinyl roof and white leather seats. Unique cast-bronze medallions identifying this 460-powered T-Bird as the "Millionth Thunderbird, 1955–1972" were added to the landau bars. And it was the only T-Bird produced in 1972 with white bodyside moldings.

After serving temporary duty as a prize for the Best of Show winner at the Classic Thunderbird Club International's 1972 convention, the one-millionth Thunderbird was sold to Californian George Watts, the same man who still owns the 1955 T-Bird that carries serial number 0005. In 1985, Watts sold the car to its present owner, Bob Peterson, of Cedar Rapids, Iowa.

The Thunderbird's 20th birthday arrived three years after the one-millionth T-Bird came and went, and Ford apparently had exhausted all its celebratory efforts in June 1972. No specific 20th Anniversary model was concocted; basically all 1975 Thunderbirds were considered commemorative editions. Two new appearance options were offered midyear for 1975, but such status-conscious paint and trim packages were nothing new for the biggest 'Birds, which grew even bigger that year as overall length stretched to 225.6 inches. Joining the 1974 Gold and White luxury group options were 1975's Silver and Copper luxury trim groups. Included in these deals were special half-vinyl roofs, seats done in either leather or crushed velour, deluxe wheelcovers, and specially appointed trunks.

The following year was the last for the Continental-based Thunderbirds, and thankfully so in some minds. Rising fuel costs were enough to convince many customers that bigger no longer meant better. Still others were becoming turned off by the overstuffed 'Bird's overdone opulence. In a February 1976 *Motor Trend* article entitled "Farewell to the Big Bird Thunderbird," Tony Swan called the 1976 T-Bird "a Rubenesque leather-padded salon on wheels for middle-class Sybarites."

This commemorative medallion appeared on the millionth Thunderbird's landau bars. Another such medallion was mounted inside on the dash. *Bob Peterson*

All Thunderbirds built in 1975 were considered Twentieth-Anniversary models. Two special trim options—Silver Luxury and Copper Luxury—were also introduced midyear to help mark the moment. A half-vinyl roof complemented the paint finish, as did the windshield—the glass on this Copper Luxury 1975 'Bird is tinted to match exterior color.

Before the last big 'Bird could say farewell, another limited-edition package was created, this one every bit as secretively as Abercrombie & Fitch's blue babies of 1967. In late May 1976, a total of 32 "commemorative" Thunderbirds rolled off the Wixom line. Built for dealer promotions in Charleston, West Virginia, and the Cincinnati-Dayton area, these cars featured silver metallic paint with contrasting black vinyl half-roofs and black leather interiors. Additionally, the American Sunroof Corporation added a special moonroof and a simulated spare tire mount atop the rear deck. Not much is known about these cars; not even Ford officials can answer questions about them since they were promotional products inspired from outside the company, not from within.

Thunderbirds built between 1972 and 1976 were as plush on the inside as they were big all over. Here, the Copper Luxury trim treatments carry over into the passenger compartment as well.

The bigger-must-be-better trend reversed itself after 1976. By then, huge, heavy cars with equally huge appetites for fuel and heavy bottom lines would no longer cut it in the American market. The Thunderbird was no exception. Just 10 years before, growth had been the key to the breed's continued success. In 1977, the opposite became true in a big way.

The debut of the eighth generation represented the first time a new Thunderbird rolled out in smaller form than the car it replaced. At 114 inches, the 1977 T-Bird's wheelbase was 6 inches shorter than the 1976 showboat's. Overall length dropped a whopping 17 inches, while weight went from about 5,000 pounds to around 4,000. Engine size went down, too–a 302-cid small-block V-8 came under that long hood unless a customer paid more for a 351 or 400.

Though still appearing reasonably large to the eye, the latest Thunderbird was now a mid-sized model clearly based on Ford's LTD II platform. It also now wore a midsized window sticker coming in at about $2,700 less than 1976's prices. Therein was the key to the all-new 1977 Thunderbird's success.

Easier to buy, easier to keep fueled, easier to handle, even easier on the eye, the eighth-generation Thunderbird soared where no T-Bird had gone before. The previous sales record, 92,843 in 1960,

Ford marked its 75th year in 1978 with this special Diamond Jubilee Thunderbird model, which featured a long list of standard dress-up touches like a half-vinyl roof and turbine-style aluminum wheels. Exterior shades were either Diamond Blue Metallic or Ember Metallic.

wasn't just broken, it was obliterated. Production in 1977 hit 318,140, a sixfold increase over 1976's final tally. Thunderbird sales reached their zenith the following year, finishing at 352,751. Results for 1979, though down to "only" 284,141, still stand at almost twice the next highest model-year for sales.

Highlights from this incredibly successful three-year run included the introduction of a sporty "T-roof" option in the spring of 1978. That year also brought a special Diamond Jubilee model to help commemorate Ford's 75th birthday. With a price tag of $10,106, the 1978 Diamond Jubilee T-Bird was called "the most exclusive Thunderbird you can buy." An extensive list of special touches included a monochromatic exterior done in either diamond blue metallic or ember metallic paint, a matching padded vinyl roof, unique trim, and color-keyed aluminum "turbine" wheels. Inside was a leather-wrapped steering wheel, a

A suitably plush interior featuring unique "Biscuit" cloth upholstery typically complemented the Diamond Jubilee Thunderbird's exterior. The instrument panel and steering wheel both included leather coverings.

leather-covered dash, simulated ebony wood treatments, and a 22-karat gold-plated owner's nameplate, to name just a few of the special features.

In 1979 a similar topline package, the Heritage Edition, was offered. Again, the exterior was monochromatic with matching turbine wheels. Color choices were maroon or light blue. The quickest way to identify a Heritage Edition T-Bird in 1979 was by the absence of rear quarter windows. Many of the same Diamond Jubilee trim treatments and such were also included in the Heritage Edition's equally long list of standard features. As in 1978, owners in 1979 could also put their initials on one of these prestigious Thunderbirds by way of an exclusive plaque.

All this glitter and pizzazz demonstrated that the Thunderbird legacy remained primarily one of luxury, even if the car itself had been dropped down out of Ford Motor Company's most posh ranks into the reach of the masses. Sure, the 1977–79 models were gussied-up LTD IIs. But they were still Thunderbirds.

Diamond Jubilee treatments even extended to the car's trunk.

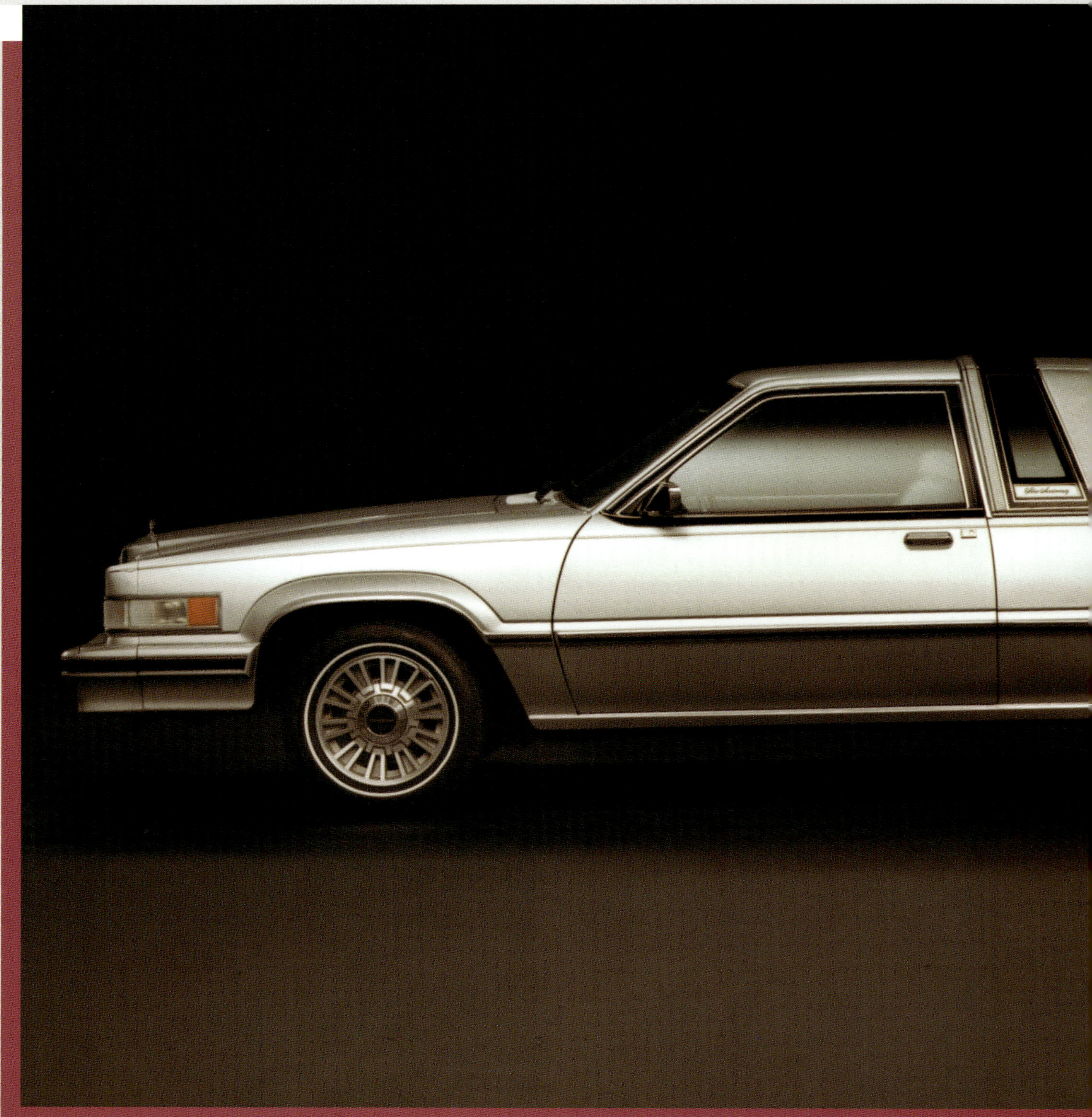

Chapter 4

Bye-Bye, Birdie—for Now

1980-1997

By 1980 Ford had sold more than 2 million Thunderbirds in a wide array of shapes and sizes: sporty two-seater, comfortable four-place cruiser, luxurious showboat with doors for as many as four, full-blown cruise ship, and midsized mass-market marvel. These were Thunderbirds all, vast disparities notwithstanding. With a new decade, however, came a new continuity. While the T-Bird shape would change three more times before the string ran out in 1997, the size would remain similar for the next 2 million 'Birds to follow after 1980.

More downsizing arrived that year, coming at a time when Ford was marketing the Thunderbird through its luxury

A redesigned, downsized Thunderbird emerged to mark the breed's quarter-century anniversary in 1980. Specifically commemorating 25 years of T-Birds was the Silver Anniversary model, introduced that spring. As before, a padded half-vinyl roof once again crowned the Thunderbird's latest flagship. *Ford Motor Company*

Following hot on the heels of the disappointing ninth-generation Thunderbird was Jack Telnack's aero coupe, introduced for 1983. This thoroughly modern bodyshell proved to be a real winner, especially so on NASCAR racetracks. Shown here is the upscale Heritage rendition of the 1983 Thunderbird. Special identification and a more plush interior marked the Heritage model.

image. However, a truly new and youthful impression was around the corner for the more mechanically-conscious sports coupes of 1983–88 and 1989–97. Differences did set the last three Thunderbird generations apart, but the trio shared the same targeted market–family types who like to ride head and shoulders above the mundane in a flashy, easy-to-handle midsized car wearing a not-so-flashy midsized price tag.

The same market pressures that in 1977 had reversed the Thunderbird's rise into Ford's elite luxury stratosphere remained strong as the 1980s dawned. A new accountability was required. Buyers needed affordable, practical cars that did more than just wow the neighbors and look impressive parked in front of the country club. Pizzazz, prestige, and performance were still valued, but so too was fuel economy, convenience, and comfort–that meant a place for the kids, too.

Readjusted thinking at Ford helped usher in the latest redesigned Thunderbird in 1980. It was based on the Fox-chassis platform, itself introduced beneath Ford's Fairmont and Mercury's Zephyr in 1978. Choosing the Fox platform meant a return to unitized body/frame construction for the Thunderbird after rolling on a full frame for 13 years.

In Thunderbird garb, the Fox chassis was stretched slightly, yet the 1980 T-Bird still measured 17 inches shorter than its 1977–79 forerunner and it weighed 700 pounds less. The wheelbase was now 108.4 inches and total length was 200.4 inches. The car was compact-looking to say the least, an impression enhanced even further by Jack Telnack's all-new angular, boxy styling. Although definitely crisp and fresh when first seen, the look quickly turned sour in naysayers' minds. As did the entire car.

In many opinions, the 1980–82 years are called the least memorable of the Thunderbird legacy. Then again, there never has been a shortage of sharply divided subjective viewpoints in the T-Bird arena, and every model group after 1957 taken pokes for one reason or another–some

The Thunderbird Turbo Coupe was instantly identified in 1983 by its unique front fascia (with foglamps) and 14-inch aluminum wheels wearing P205/70HR performance rubber. Turbo Coupes were offered up through 1988. *Ford Motor Company*

The heart of the Turbo Coupe was a 2.3-liter overhead-cam four-cylinder with electronic fuel injection and a Garrett AIResearch T-03 turbocharger. The maximum output for this force-fed four was 145 horsepower at 5,000 rpm. *Ford Motor Company*

much more than others. Until the downsized, four-place Thunderbird came along in 1980, the easiest targets had been the big 'Birds of 1967–71 and the even bigger renditions that followed.

Critics of the 1980–82 era commonly complain about blah styling that echoed a blah moment in American automotive history. Ford wasn't alone. So many of Detroit's offerings in those days looked like bricks on wheels. If anything, the 1980–82 T-Bird was trendy. Unfortunately, the trend it followed–like that whole disco thing–didn't age well.

Making matters worse was the engineering. At the time Detroit was still having difficulties catching up with federal government demands to clean up its act. The automakers needed to lower engine emissions while at the same time maximizing efficiency to keep up with rising fuel prices. Many powerplants of the early 1980s were still suffering from leaned-out, choked-up, low-compression 1970s-style maladies since electronic computer controls had yet to arrive in force to save the day.

The 1980 Thunderbird was no exception. Its new standard V-8 was a 4.2-liter (255 cid), with the 302 small-block as an option. While both V-8s offered improved fuel economy (partly because they had less weight to lug around) compared to the relative gas-guzzlers that came before, they suffered considerably in the muscle department. Net output for the 255 was 115 horsepower; 131 for the 302.

If that wasn't enough to douse your fire, Ford even introduced an optional six-cylinder–a first for the Thunderbird family–very late in 1980. This wimpy 3.3-liter (200-cid) inline-six produced an underwhelming 88 horsepower. And to add insult to injury, in 1981 it became the T-Bird's standard engine. It was joined in 1982 by another six-holer, the 112 horsepower 3.8-liter (232-cid) V-6. The 302 V-8 was dropped that year, leaving only the optional 255 for buyers trying to remember the days when the "Thunder" in Thunderbird actually meant something.

The FILA Thunderbird was created in 1984 as part of a joint effort with Italy's FILA Sports, Incorporated, an activewear manufacturer. The car's paint, lower body black-out, monochromatic wheels and trim, and pinstriping were all part of the exclusive package. *Ford Motor Company*

Setting aside any and all opinions about styling, it was what came beneath the skin of the 1980–82 T-Bird that probably sealed its fate. Engineering aspects alone left critics shaking their heads–and buyers closing their checkbooks. Thunderbird sales looked fine in 1980, even though they finished well short of 1979's sky-high results. But after 156,803 1980 'Birds rolled off dealership lots, only 86,693 were sold in 1981. The story got even worse the following year when the total for 1982 Thunderbirds fell to 45,142.

At least this generation was able to record one major claim to fame. Its first year, of course, was the Thunderbird's 25th on the road. And Ford couldn't let that moment pass without putting together a special Silver Anniversary model. Exclusive Silver Anniversary paint (other solid colors and two-tone arrangements were also offered), eye-catching cast-aluminum wheels, simulated-wood dash appointments, and a leather-wrapped steering wheel were just a few of the features included. Most agreed the Silver Anniversary Thunderbird was an impressive package, but once the party was over, the T-Bird crowd found little reason to celebrate in 1981 and 1982. This may well have been one time when three years and out was probably the right choice.

Telnack's design team came to the rescue in 1983 with the "Aerobird." Actually another downsized model, although you

The highly successful tenth-generation Thunderbird hit the streets for the last time in 1988, ending a six-year stretch where production tallies never fell below 128,000. The peak year for the Aerobird was 1984 when 170,533 were sold. This 1988 'Bird is one of 147,243 built that year.

couldn't really tell at a glance, the 1983 Thunderbird was based on a revised Fox chassis with a shortened 104-inch wheelbase. Overall impressions, however, left most witnesses seeing very little difference between the 1983 and 1982 T-Birds as far as exterior dimensions were concerned–from a curbside perspective, that is. Once inside, however, rear seat passengers quickly discovered the car's reduced nature since their space was sacrificed in order to keep driver comfort up to 1980–82 standards.

The overall length and width also decreased, all as part of a fresh, new "high-tech" plan to help the updated 10th-generation T-Bird cut through the wind like no Thunderbird ever did. Slighted by some as a "jellybean," Telnack's cutting-edge body featured an exceptional 0.35 drag coefficient–great by performance car standards, even greater in a machine many looked to for hauling the kids and family dog around town.

While the aero coupe look revived the Thunderbird's sagging image, its mechanicals remained disappointing. At first. The 3.8-liter V-6 was the base engine in 1983, but it was soon rejoined by the 302 V-8, an option added midyear. Yet another midyear performance addition really got things rolling.

With its new slippery shell, the 1983 Thunderbird was a natural for the sporty crowd. Once the Turbo Coupe was introduced in April 1983, the package was complete. Sure, the heart of the Turbo Coupe was *only* a four-cylinder engine. But this four-banger was fitted with a Garrett AIResearch turbocharger that helped squeeze out an amazing 145 horsepower from 2.3 liters. Included in the Turbo Coupe deal was a five-speed manual transmission, a 3.45:1 Traction-Lok differential, and a special handling package incorporating "Quadra-shocks" out back.

"Three decades later, the T-Bird finally becomes what it started out to be," wrote *Motor Trend's* Ron Grable concerning the 1983 Turbo Coupe. Though it didn't quite qualify as a muscle car, the turbocharged T-Bird did represent a very nice compromise between luxury–could that have been *personal* luxury?–and performance.

And if you thought the 1983 version was trick, you couldn't help but love the 1987 Turbo Coupe. With dual exhausts and an intercooler, turbo-four output increased to 190 horsepower. Four-wheel discs with ABS, automatic ride control, and 16-inch wheels were also added. *Motor Trend* editors liked it so much they named the 1987 Turbo Coupe their "Car of the Year." Turbo Coupes remained popular throughout the 10th generation's six-year run, which ended in 1988.

All of Ford's aero coupe Thunderbirds from 1983 to 1988 were strong sellers, a situation not hurt in the least by the car's rapid rise to the top of NASCAR's Winston Cup racing scene. Healthy production runs ranged from a low of 121,999 in 1983 to a peak of 170,533 in 1984. Even with its replacement waiting in the wings, the 1988 T-Bird still attracted 147,868 customers. In 1985, when 151,851 models were sold, Ford once again marked a special moment, this time with a 30th-Anniversary model.

Features included Regatta blue clearcoat metallic paint, unique bodyside moldings, and an exclusive exterior tape treatment.

Considering how much drivers appreciated the 1983–88 Thunderbird, it was obvious that its successor would have to be some special car . . . and it was. Donald Petersen's team at Ford reportedly spent $1 billion trying to upstage the 10th-generation T-Bird. That cash ended up being very well spent. The last Thunderbird rendition easily stands as the breed's best all-around combination of sportiness, practicality, luxury, and performance–all at a price that might have enticed original T-Bird buyers to come home with two or three.

Initial plans for this car, like those for the next-generation Mustang, which was also in the works, called for a smaller, front-wheel-drive model. In the Thunderbird's case, the platform would come from the wildly popular Taurus/Sable line. Luckily Ford officials changed their minds in both cases. The planned front-drive Mustang became the Probe as Dearborn's revered pony car continued on into the 1990s with rear-wheel drive and V-8 power. As for the T-Bird, it too stayed true to its heritage even as General Motors was reintroducing their competing midsized models on the front-wheel-drive GM10 platform.

To keep the thunder rolling into a fifth decade, Ford engineers created the MN12

An even sleeker Thunderbird body was introduced for 1989 along with independent rear suspension and an optional supercharged V-6. The latter feature came along as part of the Thunderbird SC package, identified here by its lower-body cladding. *Tom Wilson, courtesy of Dobbs Publishing Group*

platform featuring good ol' meat-and-potatoes rear-wheel drive. The big difference, however, came in the way those rear-drive wheels were suspended. Like Corvettes since 1963, the startling 1989 Thunderbird featured independent rear suspension. On top of that, the sleek new body was even more aerodynamic than its Aerobird predecessor. Reportedly, some 700 hours of wind-tunnel testing went into the design, which cheated the wind with a superb 0.31 drag coefficient.

The 1989 Thunderbird was slightly wider and lower than its 1988 forerunner, but it measured 3.4 inches shorter end-to-end, even though the MN12 wheelbase went nine inches beyond the Fox chassis' hub-to-hub stretch. Moving the wheels farther apart meant both ride and rear seat room were greatly enhanced. Most critics also agreed that appearances were enhanced as well. Those long, low lines made the last-generation T-Bird look like it was flying along even as it stood at rest.

If there was one complaint, it involved power choices. A V-8 was not brought back for the 1989 Thunderbird, a fact that many long-time T-Bird buyers repeatedly reminded dealers about. Even though the standard 140 horsepower 3.8-liter V-6 was no slouch, there remained no substitute for cubic inches in most 'Bird fans' minds. A V-8 was still a V-8, and a V-6 wasn't.

But Ford officials didn't turn a deaf ear to customers' clamoring for a truly hot Thunderbird. In place of the absent V-8 model, Dearborn unveiled the 1989 Super Coupe, a certifiably sexy concoction on

Ford marked 35 years of Thunderbird history in 1990 with this anniversary SC model. Monochromatic black with blue striping and silver accents was the exclusive exterior treatment on all Thirty-fifth Anniversary T-Birds. *Ford Motor Company*

The sun went down on the Thunderbird in 1997 after 43 model runs—temporarily, that is. An all-new 'Bird is expected for the new millennium.

looks alone, what with its trendy lower body cladding, fog lamps, and 16x7 cast-aluminum wheels wearing fat Goodyear Eagle rubber.

What really got T-Bird drivers' hearts pumping, however, was the powerplant that hid behind those "SC" letters impressed into the car's lower front fascia. Making the Super Coupe so super was a force-fed 3.8 liter V-6. Supplying 12 pounds of maximum boost, the SC's Eaton supercharger instantly reminded many elder Thunderbird fans of the blown 'Birds of 1957. Only this time the power package included an intercooler and sequential multiport electronic fuel injection, too. All this techno-wizardry translated into 210 horses at the rear wheels. In turn, this meant the 1989 Super Coupe could trip the lights at the far end of the quarter-mile in about 15.8 seconds–not bad at all for a car that could also turn heads at both the grocery store and the theater.

Also included in the Super Coupe deal were four-wheel ABS disc brakes, a Traction-Lok differential, automatic ride control, and heavy-duty front and rear stabilizer bars. A five-speed manual transmission was standard, with Ford's AOD (automatic-overdrive) transmission also available. Among interior goodies were articulating sport seats (with power lumbar support) and special instrumentation.

Buyers weren't the only ones impressed. "In 1987, when we named the Thunderbird Turbo Coupe our Car of

Despite its 40-year run, Ford opted not to mark the Thunderbird's transition from 1955 to 1995 with any exclusive packaging, leaving some dealerships to do the honors with commemorative decals and striping.

the Year, we called it the highest-flying Thunderbird ever produced," announced a 1989 *Motor Trend* report. "Little did we realize as we penned those words that Ford was, at the same moment, preparing a new supercharged vehicle that would take the Thunderbird name closer to the stratosphere."

The exciting 1989 Super Coupe became the third Thunderbird to win *Motor Trend's* "Car of the Year" award. And the SC Thunderbird continued impressing drivers right up to its retirement in 1995, the year the T-Bird celebrated its 40th birthday. Alone. As was the case with the 30th year of Mustang in 1994, Ford officials chose not to release a special commemorative Thunderbird in 1995. Fortunately dealers were willing to pick up the slack, adorning some 1995 T-Birds with 40th Anniversary identification.

Five years earlier, a 35th Anniversary Thunderbird was released by Ford midyear in 1990. It was appropriately based on the SC model and featured black paint with blue striping and a contrasting titanium lower body treatment. Black leather seats and black-painted wheels were also included.

In 1991, customer complaints finally influenced Ford to revive the V-8–powered Thunderbird. A 200 horsepower 5.0-liter V-8 was made an option that year. Three years later, the venerable 5.0 liter-pushrod V-8 was replaced by an optional 205 horsepower 4.6-liter modular V-8. And the Super Coupe in 1994 got a power boost to 230 horsepower thanks to a revised Eaton supercharger with low-drag, Teflon-coated rotors. A Thunderbird facelift that year also changed appearances slightly at both ends, while the basic shape remained identical.

Dearborn's decision not to officially mark the Thunderbird's 40th anniversary in 1995 was probably a result of the plain fact that the T-Bird's days were already numbered. Although the decision to end the Thunderbird's long, legendary run–albeit temporarily–in 1997 appeared abrupt and somewhat cold-hearted, market realities made it clear in 1989 that an end was in sight. Proof began to appear once it became obvious that Ford had no plans to bring out another new Thunderbird in the 1990s, instead opting to continue production, year in, year out, of basically the same car through nine nearly identical model runs.

Then, after all those years and all those cars, Dearborn also chose not to bid the final 1997 Thunderbird a fond farewell. Not even a typical press release photo appeared. According to one Lorain, Ohio, source, the historic moment was marked only by "a quiet affair for the plant workers." Of course, Ford people chose to skip a long goodbye because they didn't want this to appear as a dead end. "The Thunderbird will return by the turn of the century," was the word from Ford by way of convenient press leaks. What form the new Thunderbird will take remains, as this is being written, the $64,000 question.

The biggest question asked by Thunderbird devotees in 1997 is, why did the existing coupe have to be grounded a few years short of the new 'Bird's introduction?

The very last Thunderbird built, shown here, rolled off Ford's Lorain assembly plant on September 4, 1997. It was then donated to the Classic Thunderbird Club International in Southern California.

The answer involves Ford plans to cut back on the number of vehicle platforms offered worldwide. It has been common knowledge for some time that Ford's rear-wheel-drive platform for the Thunderbird and Mercury Cougar wouldn't survive into the new millennium. With the rug pulled out from under it, the Thunderbird simply wasn't going to be allowed to fly on in stopgap fashion until its replacement made its debut.

Besides, by 1993 the handwriting was already on the wall as sales dropped off each year thereafter, considerably so after 1995. The T-Bird essentially drove itself into the ground after Dearborn had clearly thrown in the towel in the early 1990s. As Ford Division Public Affairs Chief Jim Bright explained, "It's tough and very costly to come out with something new every three or four years." In turn, it also became tough for customers to warm up to what was basically a slightly altered remake of the previous year's model.

But that's all "chicken or egg" stuff. Plain marketing truths were the real culprits behind the move to temporarily retire the Thunderbird badge. Product planners chose not to tinker with the T-Bird in the 1990s because they had eyes. Calling two-door coupe customers "a very fickle lot," Jim Bright pointed to declining sales as the obvious reason for the retirement. In 1977, the T-Bird's share of the total car market had been a healthy 2.9 percent. Model-year sales reached their peak the following year. Then came a rapid drop-off, a trend that not even the 1989 'Bird could turn back. The 11th generation's best model-year performance came in 1993 when 133,109 Thunderbirds were sold. By 1996, total market share had fallen to a dismal 0.9 percent.

Why the sales decline? Because the car's playing field itself was on the decline. During the Thunderbird's 1970s heyday, the midsized specialty car segment made up as much as 12.2 percent (in 1976) of the total vehicle market. By 1989, that chunk had receded to 3.9 percent, and measured only 1.9 in 1996. Thunderbird wasn't alone rolling downhill; it was also joined by Chevrolet's Monte Carlo and Pontiac's Grand Prix.

Even a blind man today can see why the two-door specialty coupe has been

Ford's final Thunderbird is identified by this data plate beneath the hood. By the time you read this, the T-Bird will already be reborn.

squeezed out of Detroit's scheme of things. "A greater number of customers are now turning to light trucks instead of cars," Bright said. "Two-door coupes used to be familiar to family types in the market for utility and versatility."

The 1997 T-Bird remained an affordable, attractive package, but it never had a chance against the ever popular modern-day pickup, and even less against its wildly popular sport-utility derivative. As the 1990s wind down, SUVs continue turning the heads of many customers who earlier might have considered the specialty coupe market.

With SUVs running away with the affections of more and more car buyers, the choice became clear. Ford simply let Chevrolet and Pontiac have the ever shrinking piece of the midsized specialty pie to themselves. Pontiac's efforts to stave off the inevitable included creating a four-door Grand Prix, a tactic Dearborn never seriously considered. "Naturally we didn't want to compete against ourselves," said Bright.

Ford of course also never considered dropping the Thunderbird nameplate, considering how much this high-flying heritage has meant to the company for 40-something years now. You may already have read about the all-new T-Bird before you picked up this little book. Or maybe you've seen advance photos of it in print.

How does it look?

High-Flying Thunderbirds

The sky has always been the limit for automotive designers, especially so in the late 1940s and 1950s when aircraft-inspired fantasies seemed to dominate Detroit's drawing boards. A prime example was the new Cadillac for 1948, a trend-setting showboat that featured those curious little tailfins atop each rear quarter. Legendary designer Frank Hershey, then working under General Motors Styling Chief Harley Earl, was the man behind those fins. The inspiration for Hershey's work? None other than Lockheed's twin-tail P-38 fighter plane of World War II.

Of course, Hershey also supervised the shaping of Ford's high-flying two-seater, a car that borrowed its well-deserved name from a Native American legend that told of a grand bird that ended drought by bringing on thunder and lightning with the flap of its great wings. By coincidence, that glorious name, "Thunderbird," had already been chosen by another group to honor its vehicles for the way they performed so majestically in the heavens.

On May 25, 1953, U.S. Air Force officials activated the 3600th Air Demonstration Team out of Lukes Air Force Base in Arizona. The mission for the 3600th, the "Thunderbirds," was to prove just how reliable and safe the USAF's jet fighters were. Examples of this new technology—Lockheed's F-80 Shooting Star (America's first operational fighter jet), Republic's F-84 Thunderjet, and North American's unbeatable F-86 Sabre—had already demonstrated how well an Air Force jet could fight. The proof came in the skies over Korea beginning in 1950. With that conflict now over, the job became one of continued readiness and demonstrating how well both planes and pilots could perform.

Forty-five years after its inception, the Thunderbirds team has shown its precision-flying prowess before more than 287 million observers in all 50 states, as well as 59 foreign countries. Those beautiful red, white, and blue aircraft have put on more than 3,400 air shows, with not one performance ever canceled for mechanical reasons.

The Thunderbirds team began using the F-16 Fighting Falcon in 1983. Today they fly the upgraded F-16C. *Kevin J. Gruenwald, courtesy Thunderbirds USAF Air Demonstration Squadron, Nellis Air Force Base, Nevada*

Today's Thunderbirds fly the USAF's most versatile fighter, the General Dynamics F-16C Fighting Falcon. Six Fighting Falcons make up the squadron, which performs in precise formations and high-speed solo exhibitions. The latter shows off the F-16's maximum "performance envelope," while multi-aircraft flights, such as the famous four-plane diamond, allow the pilots to parade their exemplary discipline and meticulous training. Lasting about one hour and fifteen minutes, a typical Thunderbirds show features its most active pilot making approximately 30 maneuvers. The unit performs no more than 88 air demonstrations a year between March and November. The winter months are used for new pilot training.

The Thunderbirds Air Combat Command unit actually includes 8 pilots, 4 support officers, 4 civilians, and 130 enlisted people. About half of all personnel are replaced each year. Officers serve a two-year assignment and enlisted members serve three to four years.

Objectives for all these men and women are clear. They include: supporting USAF recruiting and retention programs; reinforcing public confidence in the Air Force and demonstrating the professional competence of Air Force crews; strengthening morale and esprit de corps among Air Force personnel; supporting Air Force community relations; representing the United States and its armed forces to foreign nations; and projecting international goodwill. In 1996, the Thunderbirds squadron did a nine-country, 30-day deployment in Europe. Included were the former communist countries of Romania, Slovenia, and Bulgaria. After returning stateside, the team did a flyover to help open the Atlanta Olympics in July. An estimated 3.8 billion people witnessed that televised performance worldwide.

Thunderbirds performances have featured various Air Force jets over the years, beginning with the straight-winged F-84G Thunderjet in 1953. Republic's improved swept-wing F-84F Thunderstreak became the jet of choice early in 1955. More than 9 million spectators at 222 air shows in North and South America saw the F-84 perform in red, white, and blue regalia.

In 1956, the Thunderbirds team moved to its present home at Nellis Air Force Base in Nevada. That year the team also traded the F-84 for North American's F-100 Super Sabre, the world's first supersonic fighter. More than 1,100 demonstrations were flown using F-100C and F-100D models up into 1969. For six brief shows in 1964, the Thunderbirds squadron used Republic's F-105B Thunderchief before turning to the F-100D.

From 1969 to 1973, the team used the Air Force's latest frontline fighter, the McDonnell-Douglas F-4E Phantom II. Phantoms were featured at more than 500 air shows. Then, to cut costs, the Thunderbirds in 1974 traded the big F-4 for the much smaller, more fuel-efficient Northrop T-38A Talon—the world's first supersonic jet trainer. A two-seat variant of Northrop's popular F-5 Freedom Fighter, the T-38 flew almost 600 demonstrations for the Thunderbirds team, including a flyover at the opening of the National Air and Space Museum in Washington, D.C., in 1976.

Early in 1983, the Thunderbirds squadron returned to "frontline status," dropping the little Talon in favor of the F-16A Fighting Falcon. More than 16.5 million people in 33 states saw the Thunderbirds fly the F-16 that first year. Additional memorable performances included a flyby during the Statue of Liberty rededication on the Fourth of July in 1986 and a trip to China in 1987, which was the first time an American demonstration squadron performed in a communist country. Five years later, the team upgraded to the F-16C.

To see the Thunderbirds perform in the air is as easy as contacting an Air Force base near you for the year's schedule. To see nostalgic examples of former Thunderbirds performers, you might consider visiting the Octave Chanute Aerospace Museum, located on the grounds of the defunct Chanute AFB in Rantoul, Illinois. There you'll find red, white, and blue representatives of both the straight- and swept-wing F-84, the F-100 Super Sabre, the T-38 Talon, the F-16, and even the F-105 "Thud." Only the F-4 is not represented.

For information about the Octave Chanute Aerospace Museum itself, located at 1011 Pacesetter Drive in Rantoul, call (217) 893-1613. Museum hours are 10 A.M. to 5 P.M. weekdays (closed on Tuesdays), Saturday 10 A.M. to 6 P.M., and Sunday noon to 5 P.M. Thunderbird fans are always welcome.

98
PURE
EQUIPPED WITH
CHAMPION
SPARK PLUGS

Chapter 5

Battling 'Birds: A Proud Competition Heritage

Ford's decision to shoot down the Thunderbird in 1997 did more than just disappoint a loyal following on Mainstreet U.S.A. It also left NASCAR's various Blue Oval racing teams wondering where they were going to find a ride.

The midsized T-Bird had carried the Ford banner proudly around stock-car racing's premier circuit since 1978. And it did so especially well in the 1980s in the hands of "Million Dollar" Bill Elliott, who set speed records, won races, and took home winnings like nobody's business. In the 1990s it was underdog Alan Kulwicki and young turk Davey Allison who stepped up to keep the Thunderbird flying high in NASCAR competition.

Extensively lightened through the use of aluminum panels and the deletion of all nonessential items, this not-so-polite two-seater certainly was ready for action, thus its name—"Battlebird." Two such Battlebirds were specially prepared for Daytona Beach's annual Speed Weeks trials. The other car did not survive. Notice the clear aerodynamic covers over the headlights.

Ford originally shipped this T-Bird from Dearborn to DePaolo Engineering in Long Beach, California, in December 1956. There it was transformed from personal luxury cruiser into all-out race car. The aluminum tailfin represented state-of-the-art streamlining for 1957, as did the small plexiglass windscreen.

Tragedy struck in 1993 when both Ford Motorsports and the entire racing community lost Kulwicki and Allison as each was killed off the track in aviation accidents. They're still dearly missed.

Mark Martin then emerged as Ford's top NASCAR pilot, joined by former Pontiac man Rusty Wallace in 1994. Two years later, Dale Jarrett made his way to the front of the pack. In 1997, Jarrett and Martin kept their T-Birds in the hunt for the Winston Cup championship right up to the season's last race, the NAPA 500 at Atlanta Motor Speedway. The NAPA 500, however, also represented the NASCAR Thunderbird's final fling. And so ended a competition legacy that had produced 184 NASCAR victories since 1978.

In July 1997, Ford finally squelched many months' worth of speculation by announcing that the lame duck Thunderbird would be replaced on the 1998 NASCAR scene by, of all things, the Taurus. A four-door NASCAR racer? You got it. And a highly successful one at that. Despite feeble early test results (probably due to sandbagging tactics), NASCAR's first four-door stock car hit the ground running, taking so many top places in the early 1998 races that rules moguls were forced to try to reshape Ford's widely perceived advantage with some mandated rear spoiler adjustments. These changes achieved little. Much of the competition probably wishes the Thunderbird had never left.

The Battlebird's heart was a stroked Y-block, a 348-cube screamer fed by Hilborn injectors. The black stenciled figure on the passenger's side fenderwell represents the last three digits of the car's serial number. This identification is found in various places throughout this factory racing machine.

The Battlebird's interior was completely stripped and refitted with no-nonsense competition gear. Notice the cooling louvers punched into the passenger-side door.

That's not to say the T-Bird was less of a battler on the track. The modern Taurus shell–translated into NASCAR specifications–is just so much better equipped to cheat the wind at superspeedway velocity. Thunderbirds over the years have always been able to ruffle rivals' feathers at competition venues of all kinds, from the hard-packed sands of Daytona Beach and the dangerously loose gravel of Pikes Peak, to the curves of sports car racing's legendary road courses and the blinding salt at Bonneville. Quarter-mile tracks also have been scorched by more than one T-Bird since 1955, with veteran Ford drag racer Bob Glidden turning most heads at the wheel of his Pro Stock 'Bird in the 1980s.

A case of split personality? During their 40-odd years on the street, Thunderbirds were personally prestigious and often luxurious by nature. Yet, at times, they were clearly no strangers to the brutal world of flat-out racing as well. The 1960s and much of the 1970s don't count. Before and after, however, Thunderbird was more than just a name.

Making the jump from mild-mannered motoring to speed-crazed competition in the beginning was no big deal for the two-seat 'Bird. While Ford did play down the sporting side of the original Thunderbird's nature, owners with a real need for speed couldn't overlook the performance potential of such a small car with such a big (for the time) and powerful V-8. As early as

Racing Thunderbirds were on hand for the first running of the Daytona 500 in February 1959. Fritz Wilson's "Museum of Speed" T-Bird has since been fully restored and now resides in the Klassix Auto Museum, located just west of Daytona International Speedway.

February 1955, Thunderbirds were showing up at sports car racing events like the then-young 12-hour endurance classic at Sebring, Florida. Other battle-ready 'Birds also started appearing at regional roundy-round events in 1955 and at newly founded NHRA dragstrips in 1956.

Pure, straight-line speed was the early 'Bird's main claim to fame in the competition arena. Remember, despite being commonly matched up against the Corvette, the two-seat T-Bird was no sports car. No way, no how. Ford people themselves weren't shy about admitting that plain truth, although they had no qualms about announcing Thunderbird speed records as part of their 1956 ad campaign.

Inspiration for those advertisements had come at Daytona Beach that February. So what if Ford's two-seater couldn't match America's sports car on a twisting road course? Its inherently high power-to-weight ratio meant it could head for the top end in every bit as much of a hurry as the Corvette . . . with the right pieces under the hood, that is.

Thunderbirds returned to NASCAR racing in 1978 after an 18-year hiatus. Here, Bobby Allison pilots his Bud Moore 'Bird at Bristol International Raceway in Bristol, Tennessee, in April 1979. Allison finished second that day in the Southeastern 500. *Dorsey Patrick*

In 1956, those underhood "adjustments" came courtesy of speed merchants like Chuck Daigh, who took his overbored, streamlined Thunderbird to Florida for NASCAR's Speed Week trials. His main competition there was Chevrolet's Corvette team led by Zora Arkus-Duntov. Daigh's T-Bird initially ran faster than Duntov's 'Vette before both cars were disqualified for being too overbored. Daigh returned to the beach with a stock-spec 312 V-8 fed by dual four-barrel carbs and recorded a 88.779-miles-per-hour average for the standing-mile. In the flying-mile, an Andy Hotten–built (he of Dearborn Steel Tubing) Thunderbird ran 134 miles per hour, good for third place. Even before the Atlantic coast sands had settled, Ford hype-masters had already changed the name to "Thunder Beach."

A truly serious Thunderbird team returned to Daytona in 1957 with full, unabashed factory backing from Dearborn. This team featured four cars, each specially prepared by DePaolo Engineering in Long Beach, California. Two were kept relatively stock in appearance, although they were powered by stroked versions of Ford's optional supercharged Y-block V-8. The other pair were treated to major modifications inside and out. These full-blown experimental racing machines were soon known as "Battlebirds."

The Battlebirds were, in 1957 terms, state-of-the-art factory race cars. At DePaolo Engineering, their frames and suspensions were beefed up and the whole package lightened by eliminating the bumpers and trading the stock doors, hoods, trunks, and headlight rims for pieces hammered out of aluminum. One Battlebird was fitted with a big, stroked 430 cid Lincoln V-8. The other was powered by a 312 Y-block stroked to 348 cubic inches.

While the Lincoln-powered Battlebird was later destroyed, its running mate managed to survive the rigors of

Almost forgotten after his long string of championships at the wheel of Chevrolet products is Dale Earnhardt's short career in a NASCAR Thunderbird. "The Intimidator" raced for Bud Moore in 1982 and 1983. Earnhardt and his T-Bird are shown here on the way to finishing fourth in the Winston Western 500, run at Southern California's Riverside International Raceway on November 20, 1983. *Phillip Salazar*

racing, as well as the more typical ravages of time, and presently resides in Bo Cheadle's fabulous Ford performance collection in California. In the early 1990s it was lovingly restored by classic Thunderbird expert Gil Baumgartner, who painstakingly researched every nut and bolt while re-creating the way the #98 Battlebird looked when it took to the beach in February 1957.

Streamlined touches included plexiglas headlight covers, a cutdown racing windscreen, and a big "Buck Rogers" fin running down the rear deck. Beneath the aluminum hood, the 348-cid Y-block was mounted six inches farther back than stock for better weight balance. Headers were added, as was Hilborn fuel injection and a Vertex magneto in place of the stock distributor. A supercharger was also tried along with the injectors at Daytona in 1957.

Since Ford didn't have a four-speed manual transmission for the 1957 Thunderbird, a Jaguar close-ratio four-speed was adapted to the Battlebird's Y-block. A

The sleek Aerobird body also made the Thunderbird a suitable candidate for SCCA road racing, as John Bauer demonstrates in Trans-Am competition at Riverside in September 1983. *Phillip Salazar*

Bill Elliott became "Million Dollar Bill" in 1985 by copping NASCAR's first "Winston Million," an R. J. Reynolds-sponsored cash bonus awarded to the winner of three of the "Big Four" superspeedway races: the Daytona 500, Talladega's Winston 500, Charlotte's World 600, and Darlington's Southern 500. Elliott's sensational 1985 season included 11 victories in all, yet he and his Thunderbird still finished second to Darrell Waltrip for the championship. *Dorsey Patrick*

Halibrand quick-change rear end in back typically allowed a varying choice of final-drive ratios. And horsepower was finally transferred into speed by way of four Halibrand magnesium knock-off wheels.

Cheadle's Battlebird was driven by Chuck Daigh at Daytona in 1957. Danny Eames got the job of piloting its Lincoln-powered running mate. On the beach, Daigh managed an amazing 205-mile-per-hour flying-mile in his brutal 'Bird but couldn't make the mandatory return pass due to engine problems–a two-way average speed was required to make the record books. Thus, Daigh's impressive one-way run ended up as nothing more than a Speed Week footnote.

At least Eames' Battlebird did take the official flying-mile title (in the Sports Class B-Modified class), averaging 160.356 for the two-way run, better by 30 miles per hour than the second-place

Cadillac-powered Corvette. Eames also scored the best standing-mile acceleration run (for experimental cars) at 98.065 miles per hour. Daigh finished third in this event with 93.065 miles per hour. As for the two other DePaolo Engineering T-Birds, they ran 1-2 in the stock-class sports car flying-mile at 138.755 and 135.313 miles per hour. Again, these speeds topped the competition by a wide margin.

At that point, the sky appeared to be the limit for the Thunderbird. But then came the Automobile Manufacturers Association's so-called ban on factory racing involvement. In the summer of 1957 that decree convinced Ford Division chief Robert McNamara to slam the engineering department's back door shut on such high-powered projects as the Battlebirds. The two beastly T-Birds quickly found their way into private hands, then rolled

Dominating NHRA Winston Pro Stock pilot Bob Glidden chose Ford's new Probe over the redesigned Thunderbird in 1989, primarily due to a perceived disadvantage created by the 1989 'Bird's longer wheelbase. While driving the shorter Aerobird in 1988 (shown here), Glidden romped and stomped his way to his ninth Winston Pro Stock title. He would win his 10th—and fifth straight—with the Probe in 1989. *Donald Farr, courtesy Super Ford magazine, Dobbs Publishing Group*

Leonard Vahsholtz started racing Thunderbirds up Colorado's famed Pikes Peak in 1984 and never looked back, taking the legendary hillclimb's stock-class title in 1986, 1987, and 1988. *Bob Jackson*

into obscurity until old #98 re-emerged some 35 years later, courtesy of Cheadle and Baumgartner.

With Ford "officially" out of sanctioned competition after 1957, it was left to Holman & Moody in Charlotte, North Carolina, to keep the thunder rolling. Soon recognized as Dearborn's unofficial racing wing, the Holman & Moody firm was requested by Ford in 1959 to build a new breed of track-ready T-Birds, this one for NASCAR's stock-car circuit. Six of these beefed-up 'Birds, armed with 350-horsepower 430 Lincoln V-8s, showed up at Bill France's new Daytona International Speedway in February 1959 for the first running of the Daytona 500. One of these cars, Fritz Wilson's #64 "Museum of Speed" entry, survived its competition career and was later restored. Today, it is on display at Daytona's Klassix Auto Museum just west of the Speedway.

On race day in 1959, Johnny Beauchamp's #73 Thunderbird raced to the wire with Lee Petty's Oldsmobile, and was initially called the winner of the first

Thunderbirds have also made their presence felt in drag racing's top fuel ranks. Here, Eric Reed's Funny Car T-Bird lights 'em up at the Mid-South Nationals in 1989. *Bob Plumer, courtesy Drag Racing Memories, Highland Springs, Virginia*

Daytona 500. But after further review, finish-line photography showed Petty's car had actually taken the checkered flag by about two feet, leaving Beauchamp to snatch defeat from the jaws of victory. Two Thunderbirds did later make their way into NASCAR winners' circles that year.

Racing T-Birds, both 1959 leftovers and a few new models, returned to NASCAR tracks in 1960, but none won a race. The arrival of the Roundbird in 1961 signaled a temporary end to the Thunderbird's short, early career in big-time stock-car racing.

That career was revived in 1978, and in glorious fashion to boot. This time cameras weren't required as Bobby Allison drove his #15 Bud Moore Thunderbird first

In 1989, Lyn St. James drove this specially modified Thunderbird to a women's closed speed record of 212.577 miles per hour at Talladega. Power came from an Ernie Elliott-built 377-cubic-inch V-8. *Donald Farr, courtesy Super Ford magazine, Dobbs Publishing Group*

across the line at the Daytona 500 in February on the way to a second-place finish in the Winston Cup Grand National seasonal points race. Obviously, Ford's downsizing of the Thunderbird in 1977 had transformed the proud 'Bird back into a candidate for competition. Unfortunately, continued downsizing in 1980 only put the NASCAR effort in reverse. True speedway glory didn't come until the bricklike 1980–82 Thunderbird was replaced by the sleek Aerobirds in 1983. Bill Elliott scored his first Winston Cup victory that year in a wind-cheating Thunderbird.

Two years later, Elliott and his T-Bird really got hot, taking a record 11 superspeedway wins, including four in a row. Awesome Bill from Dawsonville also gained a second clever nickname in 1985 by becoming the first Winston Million winner. Million Dollar Bill earned the big prize by winning three of NASCAR's "Big Four" races–the Daytona 500, Talladega 500, and Southern 500. All told, Elliott raked in $2,433,187 in earnings that year, another record.

In 1986, he established a speed standard–212 miles per hour during Talladega 500 qualifying–that still stands, and probably will forever considering the effort NASCAR rules moguls have since made to limit such astronomical top ends. Elliott's record-breaking efforts were finally rewarded with a Winston Cup championship in 1988, the fourth for a Ford driver.

The Thunderbird's final farewell came in November 1997 at Atlanta's NAPA 500. While Dale Jarrett (number 88) and Mark Martin were running their T-Birds hard for the title, Jeff Gordon (number 24) conservatively cruised his Chevrolet to the finish for NASCAR's Winston Cup championship.

Alan Kulwicki made it five in 1992, the same year Ford copped its first manufacturer's title. Thunderbirds were dominant from the start, winning the 1992 season's first nine races–four in a row by Elliott. A more consistent Kulwicki (he had but two wins) finished only 10 points ahead of Awesome Bill for the driver's title in the closest championship run to date.

Thunderbird dominance resurfaced in 1994 as Ford pilots captured 20 races, with Pontiac refugee Rusty Wallace taking home eight trophies in his new T-Bird. Mark Martin scored half of Ford's eight wins in 1995, one of which was Dale Jarrett's first for Robert Yates Racing.

Jarrett's Thunderbird roared to a third-place finish in 1996 Winston Cup seasonal points. Then he and Martin ran hard all year at the top in 1997, eventually falling just short. Needing an eighteenth-place finish at the season-ending NAPA 500 in Atlanta to clinch the championship, Chevrolet driver Jeff Gordon cruised to seventeenth place, while Jarrett and Martin battled to second and third, respectively, behind Bobby Labonte's winning Pontiac. Gordon captured his third Winston Cup title in three years with 4,710 points. Jarrett was second with 4,696, and Martin third at 4,681. It was NASCAR's fourth closest victory margin, and the tightest ever measured from first to third.

It was also a suitably dramatic send-off for Ford's NASCAR Thunderbird, as high-flying a competitor as stock-car racing has ever seen.

Index